Application-Specific Mesh-based Heterogeneous FPGA Architectures

Husain Parvez • Habib Mehrez

Application-Specific Mesh-based Heterogeneous FPGA Architectures

Springer

Husain Parvez
Université Pierre et Marie Curie
Paris VI, Laboratoire LIP6
Département SoC, Equipe CIAN
4, Place Jussieu
75252 Paris
France
husainahmed154@hotmail.com

Habib Mehrez
Université Pierre et Marie Curie
Paris VI, Laboratoire LIP6
Département SoC, Equipe CIAN
4, Place Jussieu
75252 Paris
France
habib.mehrez@lip6.fr

ISBN 978-1-4899-8788-4 ISBN 978-1-4419-7928-5 (eBook)
DOI 10.1007/978-1-4419-7928-5
Springer New York Dordrecht Heidelberg London

Softcover re-print of the Hardcover 1st edition 2011

Printed on acid-free paper

Springer is part of Springer Science+Business Media (www.springer.com)

Foreword

This book concerns the broad domain of reconfigurable architectures and more specifically FPGAs. Different issues that are the centre of this book are very essential and are intended to overcome the current limitations of FPGAs, which are experiencing extremely rapid and sustained development for several years. In fact, FPGAs offer a particularly remarkable flexibility but suffer from a level of performance that can be disadvantageous for some applications in terms of surface, speed or energy. This work presents several significant and original contributions in order to remove these limitations by focusing especially on surface metric.

This book aims at exploring heterogeneous FPGA architectures dedicated to a given set of application circuits. Beyond architecture exploration, this work also presents automatic FPGA "layout" generation flow, and a new component called as an ASIF "Application Specific Inflexible FPGA", which significantly reduces silicon footprint by customizing the architecture for a given set of applications circuits. The importance and originality of the contributions made in this work revolve around this new concept of application specific reconfigurable circuits, mainly the development of an entire design environment including: generation tools, floor-planning, placement and routing adapted to the case of heterogeneous blocks. Careful analysis of results and the validation of proposed techniques have also been observed.

The monograph of this book is based on Husain's doctoral thesis. It was a great pleasure for me to supervise his thesis. This book will be of special interest for students and researches in the domain of FPGA architectures in general, and application-specific FPGA architectures, heterogeneous FPGA architectures, and their automatic hardware generation in particular.

Prof. Dr. Habib MEHREZ
Team Lead, Analog and Digital Integrated Circuit Group at
System-on-Chip department of LIP6
Professor at University of Paris 6 (UPMC), Paris, France

Preface

1 Abstract

Low volume production of FPGA-based products is quite effective and economical because they are easy to design and program in shortest possible time. The generic reconfigurable resources in an FPGA can be programmed to execute a vast variety of applications at mutually exclusive times. However, the flexibility of FPGAs makes them much larger, slower, and more power consuming than their counterpart ASICs. Consequently, FPGAs are unsuitable for applications requiring high volume production, high performance or low power consumption. The main theme of this work is to reduce area of FPGAs by introducing heterogeneous hard-blocks (such as multipliers, adders etc) in FPGAs, and by designing application specific FPGAs. Automatic FPGA layout generation techniques are employed to decrease non-recurring engineering (NRE) costs and time-to-market of application specific heterogeneous FPGA architectures.

This work initially presents a new environment for the exploration of mesh-based heterogeneous FPGA architectures. An architecture description mechanism allows to define new heterogeneous blocks. A variety of automatic and manual options can be selected to optimize floor-planning of heterogeneous blocks on the FPGA architecture. The exploration environment later allows to test different benchmark circuits on the newly defined heterogeneous FPGA architecture. An automatic FPGA layout generator is presented which generates a tile-based FPGA layout for a subset of architectures generated by our exploration environment. We have successfully taped-out a 1024 Look-Up Table based mesh FPGA architecture using 130nm 6-metal layer CMOS process of ST.

The Heterogeneous FPGA exploration environment is further enhanced to explore application specific FPGAs. If a digital product is required to provide multiple functionalities at exclusive times, each distinct functionality represented by an application circuit is efficiently mapped on an FPGA. Later, the FPGA is reduced for the given set of application circuits. This reduced FPGA is proposed and termed here as an Application Specific Inflexible FPGA (ASIF). The main idea is to perform prototyping, testing and even initial shipment of a design on an FPGA; later it can be migrated to an ASIF for high volume production. ASIF generation techniques can also be employed to generate a single configurable ASIC core that can perform multiple tasks at different times. An ASIF for 20 MCNC benchmark circuits is found to be 82% smaller than a traditional mesh-based unidirectional FPGA required to map any of these circuits. An ASIF can also be reprogrammed to execute new or modified circuits, but unlike FPGAs, at a very limited scale. A new CAD flow is presented which can map new application circuits on an ASIF. An automatic ASIF hardware generator is also presented.

2 Dedication

Dedicated to my parents, and to Scamoail, Hasan, Omer and Asma.

Husain Parvez

3 Acknowledgments

This book is based on my PhD. thesis at LIP6/Universiy Pierre & Marie Curie (UPMC). I am highly indebted to Professor Habib Mehrez for supervising my thesis. I am also grateful for his continual technical advice. I am extremely thankful to Professor Guy Gogniat, and Dr. Gilles Sassatelli, for reviewing my thesis and for their valuable comments.

I am immensely grateful to Dr. Hayder Mrabet and Dr. Zied Marrakchi for providing detailed and insightful technical advice. This thesis would not had been possible without their support and guidance. I am also thankful to Dr. André Tissot and Dr. Nicolas Fel for their technical guidance in the tape-out of FPGA chip.

I would also like to thank Assosiate Professor Hassan Aboushady for motivating me to publish my PhD. work in the form of this book.

Contents

List of Figures

List of Tables

1 Introduction

1.1 Background

Field Programmable Gate Arrays (FPGAs) are reconfigurable devices that can execute variety of hardware applications. A software CAD flow transforms a hardware application to a programming bitstream, which can easily be and instantly programmed on an FPGA. This re-programmability of an FPGA device can be used to execute different hardware applications at mutually exclusive times. Similarly any errors or updates in the final product can be corrected/upgraded by simply reprogramming the FPGA. An FPGA also allows partial re-configuration i.e, only a portion of an FPGA is configured while other portions are still running. Partial re-configuration is useful in designing systems which require to adapt frequently according to run-time constraints. Compared to alternate technologies that fabricate specific application circuits on silicon, FPGA-based applications have less non recurring engineering (NRE) cost and shorter time-to-market. These advantages make FPGA-based products very effective and economical for low to medium volume production.

The flexibility and reusability of an FPGA is due to its configurable logic blocks that are interconnected through configurable routing resources. An application design can easily be mapped on these configurable resources of an FPGA by using a dedicated software flow. This application is initially synthesized into interconnected logic blocks (generally comprising of Look-Up Tables and Flip-Flops). The logic block instances of the synthesized circuit are then placed on configurable logic blocks of FPGA. The placement is done in such a way that minimum routing resources are required to interconnect them. Connections between these logic blocks are later routed using configurable routing resources. The logic blocks and routing resources are programmed by static RAMs (SRAMs) which are distributed through

H. Parvez and H. Mehrez, *Application-Specific Mesh-based Heterogeneous FPGA Architectures*,
DOI 10.1007/978-1-4419-7928-5_1, © Springer Science+Business Media, LLC 2011

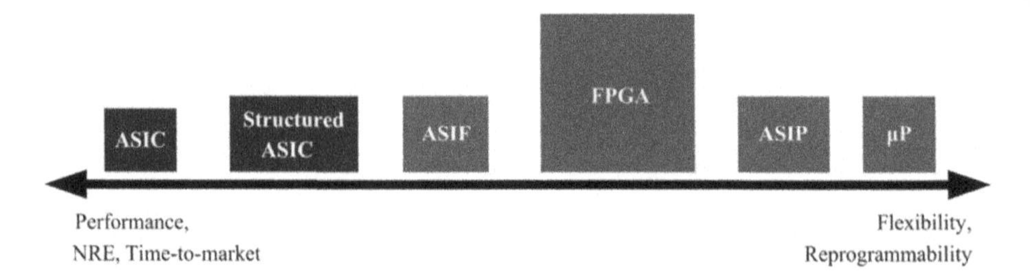

Figure 1.1: Comparison of different platforms used for implementing digital applications

out the FPGA. Once an application circuit is placed and routed on the FPGA, the SRAM bit information of the entire FPGA is gathered to form a bitstream. This bitstream is then programmed on the SRAMs of FPGA by a bitstream loader, which is integrated in the FPGA. This work focuses only on SRAM-based FPGA devices, as they are the most commonly used commercial FPGAs. Other technologies used to implement configuration memory includes antifuses [J.Greene et al., 1993] and floating gate transistors [S.Brown, 1994].

An FPGA device can be compared to other digital computational devices. The section below presents a broad overview and their comparison with FPGAs.

1.1.1 Microprocessors

A microprocessor is a general purpose hardware device that can execute a software task. The software task is represented as a stream of software instructions; each instruction belongs to a pre-defined instruction-set. The instructions in a task are executed on the microprocessor in a serial manner. These instructions are stored in instruction memory, whereas data to be processed is stored in data memory or general purpose registers. Each instruction is executed as a set of machine operations, which generally comprise of (i) Instruction Fetch (IF): A software instruction is fetched from memory. (ii) Instruction Decode(ID): Instruction is decoded, and registers are accessed from the register file. (iii) Execute (EX): Instruction is executed according to its type. The execution may include an ALU operation on operands (registers and/or immediate values). (iv) Memory Access (MEM): Loads data from memory to register, or stores data from register to memory. (v) Write back (WB): Computed result is written back to register. If each machine operation takes one machine cycle, then one software instruction will take 5 machine cycles to execute. However, pipelining technique can be employed by starting a new instruction on each machine cycle. In this way multiple software instructions are overlapped in execution.

An FPGA can be compared to a microprocessor, as both are flexible and reprogrammable. However, the configurability of FPGAs is fundamentally different from the programmability of a traditional microprocessor. A microprocessor can execute software applications, whereas an FPGA can execute hardware applications. The functional logic blocks in a traditional microprocessor cannot be configured for different software programs. Whereas logic blocks in an FPGA can be configured differently for each application. The limited number of non-reconfigurable functional units in a microprocessor are carefully selected by the microprocessor architect in accordance with the generic software needs. These functional units and their intercommunication is optimized to achieve very high clock frequencies. A microprocessor takes varying number of clock cycles to execute different software instructions.

Execution of a hardware application design enjoys inherent parallelism. On the contrary, a stream of software instructions are executed sequentially on a microprocessor. Hardware implementation can use any required hardware functionality to execute a particular task. Whereas a software implementation might require several software instructions to execute the same task. Due to these reasons, hardware implementation of a particular design is generally much faster than its software implementation. Although, due to internal architecture, the clock frequencies of FPGAs are very low as compared to that of microprocessors. Nevertheless, a task running on an FPGA is generally much faster than the same task running on a microprocessor.

1.1.2 Application-specific instruction-set processor (ASIPs)

An application-specific instruction-set processor (ASIP) is designed according to requirements of a particular application. The instruction-set of an ASIP is specifically designed to accelerate most commonly used complex functions required by the particular application domain. A general purpose microprocessor is generally designed to achieve maximum performance with maximum flexibility. Whereas, an ASIP reduces the flexibility to achieve better performance gains for a pre-defined application domain. ASIP examples includes DSP, media, network and other domain-specific processors.

1.1.3 Application Specific Integrated Circuits (ASICs)

An ASIC is a digital device that is customized to perform a particular task. The functionality of an ASIC is defined by using a hardware description language (HDL) such as Verilog or VHDL. After the functional verification of HDL, it is synthesized into a netlist of low-level gates called as standard-cells. The standard-cells are taken from a pre-characterized standard-cell library which comprise of gates such as 2 input and, 2 input nor etc. After the synthesis, a placement tool places the standard cells on a region representing the ASIC. The standard cells are placed in accordance to a variety of placement constraints. After the placement, routing tool uses the placement information and connects electrical connections between the standard cells. The final output is used to fabricate a physical IC of the ASIC. The standard-cell library can be used along with modern CAD systems can offer considerable performance/cost benefits. Instead of using standard-cell library, the layout can be designed in full-custom by laying out each individual transistor and their interconnection. Full-custom design improves the area and performance of the ASIC, however it is an extremely labor-intensive task, and requires higher skills. It increases the design time, increases non-recurring engineering (NRE) cost, and the time-to-market of the product.

When an FPGA is compared to an ASIC, it is revealed that all the advantages of an FPGA come with a very huge cost. A circuit implemented on an FPGA is generally 35 times larger, 4 times slower, and 14 times more power consuming than the same circuit implemented in a standard-cell based ASIC [I.Kuon and J.Rose, 2007]. Consequently, FPGAs are unsuitable for applications requiring high volume production, high performance or low power consumption. However, ASICs suffer from inflexibility, higher non-recurring engineering (NRE) cost and higher time-to-market. Thus, ASICs are only suitable if produced in high volumes. Moreover, any corrections or updates in an ASIC based product requires a re-spin, which further increases time-to-market.

1.1.4 Structured-ASICs

The NRE cost and time-to-market of an ASIC are reduced with the advent of a new breed of ASICs know as Structured-ASICs. Structured-ASICs consist of an array of optimized logic elements on lower mask layers. A desired functionality is implemented on it by making changes on few upper mask layers. The upper mask layers are used to customize connections between pre-defined logic elements on lower mask layers. Compared to cell-based ASICs, the design and manufacturing time is reduced in structured-ASIC due to its predefined metal layers. The power and clock distribution networks are predefined in structured-ASICs. The CAD tools generally cost lower and are easier to use than the cell-based ASIC tools.

Structured-ASICs are explored or manufactured by several companies [Wu and Tsai, 2004] [Okamoto et al., 2004] [Sherlekar, 2004] [eASIC, 2010]. FPGA vendors have also started giving provision to migrate FPGA based application to Structured-ASIC. The main idea is to prototype, test and even ship initial few designs on an FPGA; later it can be migrated to Structured-ASIC for high volume production [Hutton et al., 2006]. In this regard, Altera has proposed a clean migration methodology [Pistorius et al., 2007] that ensures equivalence verification between FPGA and its Structured-ASIC (known as HardCopy [HardCopy, IV]). However, migration of an FPGA based product to a Structured-ASIC supports only a single application design. HardCopy totally loses the quality of an FPGA to use the same hardware for executing multiple applications at different times. An Application Specific Inflexible FPGA (ASIF) proposed in this work retains this property and can be a possible future extension for HardCopy. Figure 1.1 presents a rough comparison of different platforms that can be used for implementing digital applications.

1.2 Motivation and contribution

The global theme of this work is to reduce the drawbacks of an FPGA when compared to an ASIC i.e, to make it viable for applications requiring high volume production, high performance or low power consumption. The main focus of this work is to reduce the area of an FPGA. Speed and power are not directly taken into consideration. Reduced area generally tend to give improvement in speed and power. Area of an FPGA is improved by employing two broad techniques

- Logic flexibility of an FPGA is reduced by inserting heterogeneous hard-blocks, such as multipliers, adders etc.

- Routing flexibility of an FPGA is reduced by proposing ASIF (Application Specific Inflexible FPGA)

Considerable amount of logic and routing resources can be saved if a particular hard-block is directly embedded into an FPGA. The types and quantities of hard-blocks in an FPGA can be decided from the application domain for which an FPGA is required. Similarly, the area of an FPGA can be decreased by reducing the routing network of an FPGA for a given set of application circuits. Such a reduced FPGA is called as an Application Specific Inflexible FPGA (ASIF). An ASIF can be used under two main scenarios. (i) A product is designed and tested on an FPGA. It can be migrated to an ASIF for high volume production. (ii) ASIF generation techniques can be used to generate a configurable ASIC core that can perform multiple tasks at different times.

The major contribution of this book includes the following :

1.2.1 Exploration environment for heterogeneous FPGA Architectures

A new software environment is proposed for the exploration of heterogeneous FPGA architectures. An architecture description mechanism is used to select different FPGA architectural parameters. New heterogeneous blocks can also be easily defined. A software CAD flow is used to map an application circuit on the newly defined FPGA architecture. The software flow not only maps the instances of a target netlist on their respective blocks in the architecture, but also refines the floor-planning of heterogeneous blocks on the architecture. This environment is used to define and optimize FPGA architectures for a particular application domain. Heterogeneous FPGA architectures are generated for a set of different benchmark circuits. Different automatic and manual floor-planning techniques are used to show the effect of floor-planning on the overall area of heterogeneous FPGAs.

The novelty of the proposed exploration environment is its ability to optimize the floor-planning of heterogeneous blocks on the FPGA architecture. The floor-planning can be optimized for a given set of application circuits. Existing exploration environments propose a pre-determined floor-planning organization. They do not consider the problem of optimizing the floor-planning automatically.

1.2.2 Automatic FPGA layout generation

An automated method is proposed for the generation of FPGA layout. The main purpose of developing a layout generator is to reduce the overall FPGA design time with limited area penalty. This generator works in two phases. In the first phase, a partial layout is generated using generic parameterized algorithms. The partial layout is generated to obtain a fast bitstream configuration mechanism, an optimized power routing and a balanced clock distribution network. In the second phase, the generator completes the remaining layout using automatic placer and router. This two-phase technique allows better maneuvering of the layout according to desired constraints. FPGA layout is generated using a symbolic standard-cell library which allows to migrate the symbolic layout to any fabrication process technology. The proposed method is validated by generating layout of a 1024 Look-Up Table based mesh FPGA architecture. The layout is successfully taped-out using 130nm 6-metal layer CMOS process of STMicroelectronics [STMicroelectronics, 2010].

Previously proposed FPGA layout generators use commercial VLSI tools. The proposed tile-based FPGA layout generator uses open-source VLSI tools, which can be adapted for possible future demands. Another major contribution is the concept of manual intervention in automatic layout generation. Manual intervention in layout generation, attained through generic layout parameters, allows to maneuver the layout according to desired constraints.

1.2.3 ASIF: Application Specific Inflexible FPGA

An Application Specific Inflexible FPGA (ASIF) is an FPGA with reduced flexibility that can implement a set of application circuits which will operate at mutually exclusive times. These circuits are efficiently placed and routed on an FPGA to minimize total routing switches required by the architecture. Later, all unused routing switches are removed from the FPGA to generate an ASIF. An ASIF for a set of 20 MCNC benchmark circuits is found to be 82% smaller than a mesh-based unidirectional FPGA required to map any of these circuits. An ASIF for a set of circuits is also compared with the sum of areas of standard-cell based ASICs of given circuits. An ASIF can also be reprogrammed to execute new or modified circuits, but unlike FPGAs, at a limited scale. A new CAD flow is proposed which can map application circuits on an ASIF.

A novel concept of application specific inflexible FPGA (ASIF) is proposed. Existing placement and routing algorithms are modified to propose efficient placement and routing techniques. Efficient placement and routing techniques are used to minimize the area of an ASIF. The concept of ASIF is also applied on heterogeneous FPGA architectures, thus proposing heterogeneous-ASIF. A new branch-and-bound based placement algorithm is proposed to map new application circuits on an ASIF.

1.2.4 Automatic ASIF hardware generation

An automated method of generating ASIF hardware is proposed. The ASIF VHDL generator is integrated with the ASIF exploration environment. By doing so, all the ASIF architectural parameters supported by the exploration environment can be automatically translated to a synthesizable VHDL netlist. The synthesized VHDL model of the ASIF is later passed to Cadence Encounter [Cadence, 2010] to generate its layout. The layout of ASIF is generated with a different technique than the automatic tile-based FPGA layout generator referred to earlier. The tile-based FPGA layout generator has its own advantages. However, one of its main disadvantages is that it is not directly linked to the architecture exploration environment. Any changes done in the architecture exploration environment are not immediately reflected in the layout generator. Since, ASIF is still in the process of exploration, we preferred to generate the VHDL of an ASIF by using different data-structures of ASIF exploration environment. The exploration environment also generates programming bitstream for each netlist used in the generation of on ASIF. The VHDL model of the generated ASIF is simulated using Synopsys [Synopsys, 2010]; bitstreams of different application circuits are individually programmed on the ASIF to execute different application circuits exclusively. The VHDL model of the ASIF is later passed to Cadence Encounter to generate layout of ASIF using 130nm 6-metal layer CMOS process of STMicroelectronics [STMicroelectronics, 2010].

1.3 Field of Applications

ASIF generation techniques can be employed for any digital product that provides multiple functionalities at exclusive times. Such a digital product may comprise of a video application, a multi-standard radio application, or any set of DSP functionalities required at different times. Suppose, a hardware circuit is required for executing various digital algorithms required for a camera. The lens of a camera sends RGB frames to the display screen of camera. The received RGB frames can also be encoded as a video or an image using their respective encoders. After the video or an image is recorded, they can be viewed on the camera screen using their respective decoders. Different quality requirements may require the camera to use different video compression techniques, such as H.264 and MPEG-4. Similarly different image compression techniques may also be used, such as JPEG and PNG. Such camera requires a set of 8 different digital circuits at exclusive times, i.e. encoder/decoder pairs for H.264, MPEG-4, JPEG and PNG. Each time a particular functionality is required, a control circuitry programs the bitstream of the desired function on FPGA. These eight digital circuits can be designed and tested on an FPGA. Later, for high volume production, the FPGA can be reduced to an ASIF for the given eight application circuits. Similarly, an ASIF can also be used after an RF ADC [Beilleau et al., 2009].

1.4 Book Organization

This dissertation is organized as follows. Chapter 2 presents the state-of-the-art of FPGA architectures and CAD tools. It also presents new architectures that are proposed to resolve different drawbacks of FPGAs. Chapter 3 presents an environment for the exploration of mesh-based heterogeneous FPGA architectures. Chapter 4 presents an automated method for the generation of a tile-based FPGA layout. Chapter 5 describes different ASIF generation techniques and their comparison with FPGAs. Chapter 6 presents ASIFs using heterogenous logic blocks, and their comparison with the sum of areas of standard-cell based ASICs of given circuits. It also presents a CAD flow to map new or modified application circuits on an ASIF. Chapter 7 presents automatic hardware generation of ASIFs, and their layout generation using Cadence Encounter. Finally Chapter 8 presents a conclusion and future work.

2

State of the art

2.1 Introduction to FPGAs

A Field programmable Gate Array (FPGA) is an integrated circuit that is designed to be configured after manufacturing. FPGAs can be used to implement any logic function that an Application Specific Integrated Circuit (ASIC) can perform. For varying requirements, a portion of FPGA can also be partially reconfigured while the rest of an FPGA is still running. Unlike other technologies, which implement hardware directly into silicon, any errors in the final FPGA-based product can be easily corrected by simply reprogramming the FPGA. Any future updates in the final product can also be easily upgraded by simply downloading a new application bitstream. The ease of programming and debugging with FPGAs decreases the overall non-recurring-engineering (NRE) costs and time-to-market of FPGA-based products.

The reconfigurability of FPGAs is due to their reconfigurable components called as logic blocks, which are interconnected by a reconfigurable routing network. There are two main routing interconnect topologies: Tree-based routing network [A.DeHon, 1999] [Marrakchi et al., 2009] and Mesh-based routing network [Betz et al., 1999]. A tree-based FPGA architecture is created by connecting logic blocks into clusters. These clusters are connected recursively to form a hierarchical structure. On contrary, a mesh-based FPGA architecture interconnects logic blocks through a 2-D mesh of routing network. Tree-based interconnect topology occupies less area than the mesh-based interconnect topology [Marrakchi, 2008], however a tree-based FPGA suffers from layout scalability problems. The layout of a mesh-based FPGA is scalable and is thus commonly used by commercial FPGA vendors such as Xilinx [Xilinx, 2010] and Altera [Altera, 2010]. This work focuses only on mesh-based routing

H. Parvez and H. Mehrez, *Application-Specific Mesh-based Heterogeneous FPGA Architectures*,
DOI 10.1007/978-1-4419-7928-5_2, © Springer Science+Business Media, LLC 2011

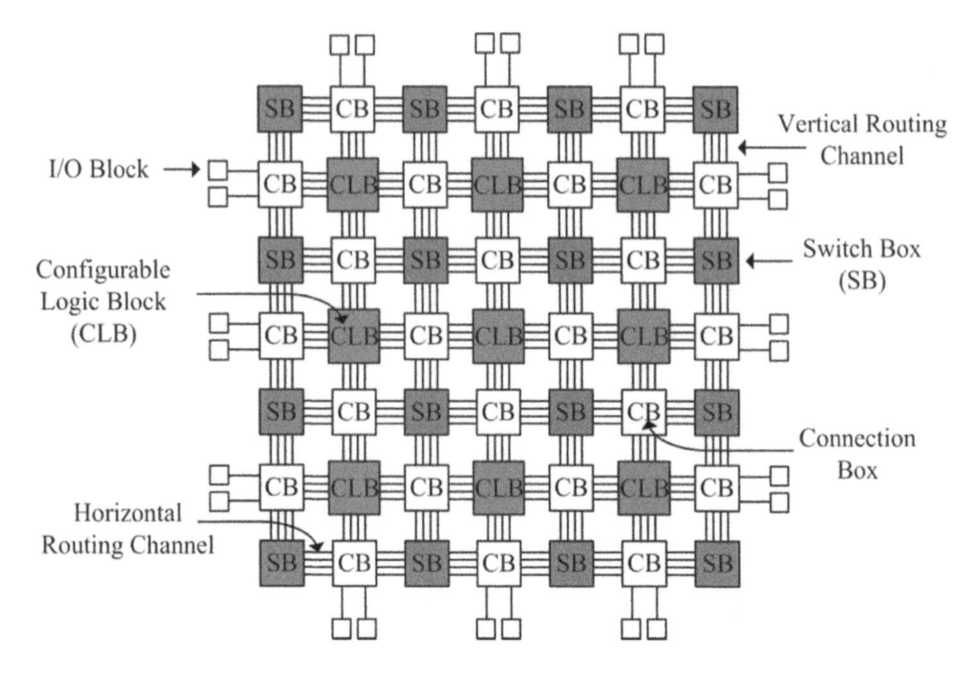

Figure 2.1: Overview of FPGA Architecture [Betz et al., 1999]

topology.

Figure 2.1 shows a traditional mesh-based FPGA architecture. The configurable logic blocks (CLBs) are arranged on a 2D grid and are interconnected by a programmable routing network. The Input/Output (I/O) blocks on the periphery of FPGA chip are also connected to the programmable routing network. The routing network comprises of horizontal and vertical routing channel tracks. Switch boxes connect horizontal and vertical routing tracks of the routing network. Connection boxes connect logic and I/O block pins with adjacent routing tracks. A software flow converts a target hardware circuit into interconnected CLBs and I/O instances, and then maps them on the FPGA. The software flow also generates a bitstream, which is programmed on the FPGA to execute the target hardware circuit. The mesh-based FPGA, and its software flow is described in detail as below.

2.1.1 Configurable Logic Block

A configurable logic block (CLB) is a basic component of an FPGA that implements logic functionality of a target application design. A CLB can comprise of a single basic logic element (BLE), or a cluster of locally interconnected BLEs. A simple BLE consists of a Look-Up Table (LUT), and a Flip-Flop. A LUT with k inputs (LUT-k) contains 2^k configuration bits; it can implement any k-input boolean function. Figure 2.2 shows a simple BLE comprising of a 4 input Look-Up Table (LUT-4) and a D-type Flip-Flop. The LUT-4 uses 16 SRAM (static ran-

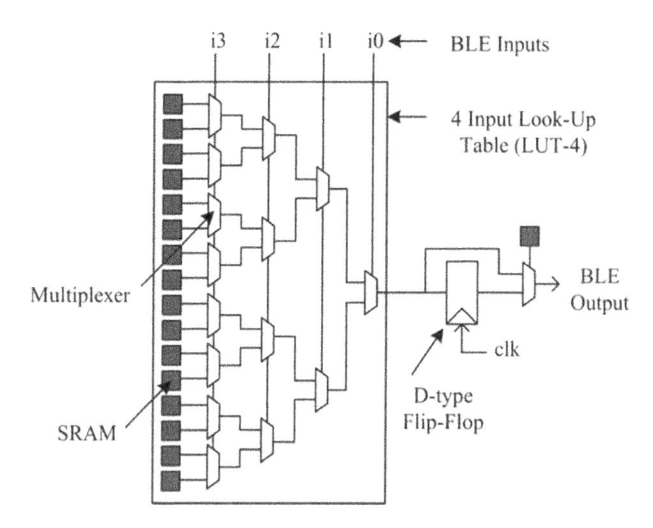

Figure 2.2: Basic Logic Element (BLE)

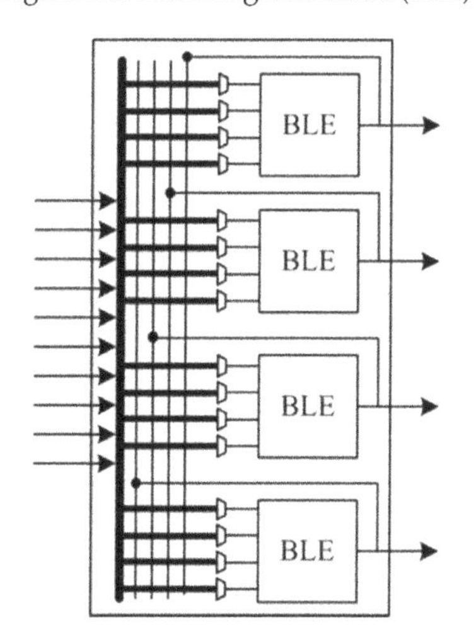

Figure 2.3: A Configurable Logic Block (CLB) having four BLEs

dom access memory) bits to implement any 4 inputs boolean function. The output of LUT-4 s connected to an optional Flip-Flop. A multiplexor selects the BLE output to be either the output of a Flip-Flop or the LUT-4.

A Look-Up Table with more number of inputs reduces the total number of LUTs required o map a hardware circuit. More logic functionality can be mapped in a single LUT. This eventually reduces the intercommunication between LUTs, and thus the speed of hardware circuit improves. However, a LUT with more number of inputs increases its area exponentially. [J.Rose et al., 1990] and [E.Ahmed and J.Rose, 2000] have measured the effect of the

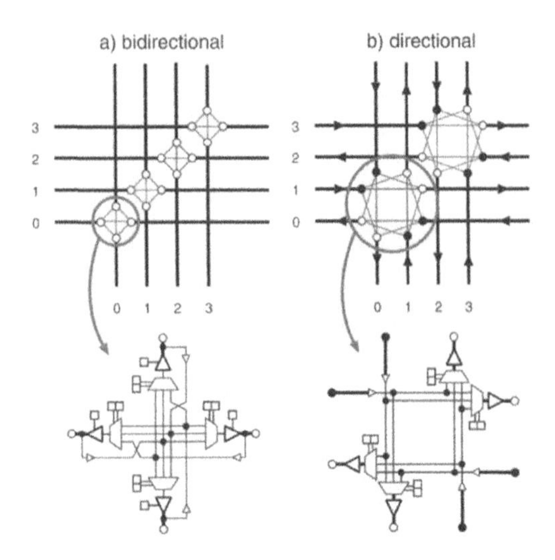

Figure 2.4: Switch Block, length 1 wires [G.Lemieux et al., 2004]

number of LUT inputs on area, speed and routability of FPGAs. They have concluded that 4-input LUTs provide a good tradeoff between speed and density of FPGAs.

A CLB can contain a cluster of BLEs connected through a local routing network. Figure 2.3 shows a cluster of 4 BLEs; each BLE contains a LUT-4 and a Flip-Flop. The BLE output is accessible to other BLEs of the same cluster through a local routing network. The number of output pins of a cluster are equal to the total number of BLEs in a cluster (with each BLE having a single output). Whereas the number of input pins of a cluster can be less than or equal to the sum of input pins required by all the BLEs in the cluster. Modern FPGAs contain typically 4 to 10 BLEs in a single cluster.

2.1.2 Routing Network

The routing network of an FPGA occupies 80-90% of FPGA chip area, whereas the logic area occupies only 10-20% area [Betz et al., 1999]. The flexibility of an FPGA is mainly dependent on its programmable routing network. A mesh-based FPGA routing network consists of horizontal and vertical routing tracks which are interconnected through switch boxes (SB). Logic blocks are connected to the routing network through connection boxes (CB). The flexibility of a connection block (Fc) is the number of routing tracks of adjacent channel which are connected to the pin of a block. The connectivity of input pins of logic blocks with the adjacent routing channel is called as Fc(in); the connectivity of output pins of the logic blocks with the adjacent routing channel is called as Fc(out). An Fc(in) equal to 1.0 means that all the tracks of adjacent routing channel are connected to the input pin of the logic block. An Fc(in) equal to 0.5 means that only 50% tracks of the adjacent routing channel are connected to the input pin.

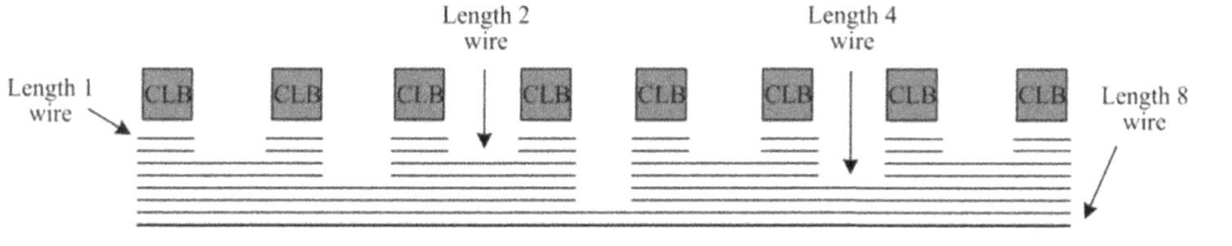

Figure 2.5: Channel segment distribution

The flexibility of switch box (Fs) is the total number of tracks with which every track entering in the switch box connects to. The routing tracks connected through a switch box can be bidirectional or unidirectional (also called as directional) tracks. Figure 2.4 shows a bidirectional and a unidirectional switch box having Fs equal to 3. The input tracks (or wires) in both these switch boxes connects to 3 other tracks of the same switch box. The only limitation of unidirectional switch box is that their routing channel width must be in multiples of 2. Multi-length wires are created to reduce area and delay. Figure 2.5 shows an example of different length wires. Longer wire segments span multiple blocks and require fewer switches, thereby reducing routing area and delay. However, they also decrease routing flexibility, which reduces the probability to route a hardware circuit successfully. Modern commercial FPGAs commonly use a combination of long and short wires to balance flexibility, area and delay of the routing network .

Generally, the output pins of a block can connect to any routing track through pass transistors. Each pass transistor forms a tristate output that can be independently turned on or off. However, single-driver wiring technique can also be used to connect output pins of a block to the adjacent routing tracks. For single-driver wiring, tristate elements cannot be used; the output of block needs to be connected to the neighboring routing network through multiplexors in the switch box. Modern commercial FPGA architectures have moved towards using single-driver, directional routing tracks. [G.Lemieux et al., 2004] show that if single-driver directional wiring is used instead of bidirectional wiring, 25% improvement in area, 9% in delay and 32% in area-delay can be achieved. All these advantages are achieved without making any major changes in the FPGA CAD flow.

2.1.3 Software Flow

One of the major research aspects of FPGAs is the development of software flow required to map hardware applications on an FPGA. The effectiveness and quality of an FPGA is largely dependent on the software flow provided with an FPGA. The software flow takes an application design description in a Hardware Description Language (HDL) and converts it to a stream of bits that is eventually programmed on the FPGA. Figure 2.6 shows a complete software flow for programming an application circuit on a mesh-based FPGA. A brief description of various modules of software flow is described below.

Logic synthesis :

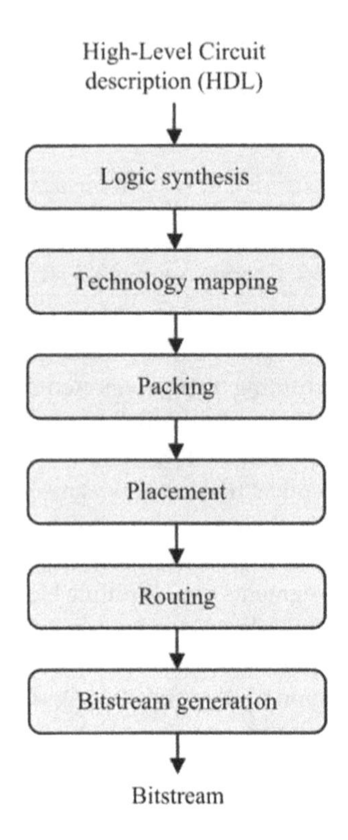

Figure 2.6: FPGA software configuration flow

Logic synthesis [Brayton and McMullen, 1982] [Brayton et al., 1990] transforms an HDL description (VHDL or Verilog) into a set of boolean gates and Flip-Flops. The synthesis tools transform the register-transfer-language (RTL) description of a design into a hierarchical boolean network. Various technology-independent techniques are applied to optimize the boolean network. The typical cost function of technology-independent optimizations is the total literal count of the factored representation of the logic function. The literal count correlates very well with the circuit area. Further details of logic synthesis are beyond the scope of this work.

Technology mapping :

After logic synthesis, technology-dependent optimizations are performed. These optimizations transform the technology-independent boolean network into a network of gates in the given technology library. The technology mapping for FPGAs transforms the given boolean network to the available set of blocks on an FPGA. For a traditional FPGA architecture, the boolean network is transformed into Look-Up Tables and Flip-Flops. Technology mapping algorithms optimize a given boolean network for a set of different objective functions including depth, area and power. The FlowMap algorithm [J.Cong and Y.Ding, 1994a] is a

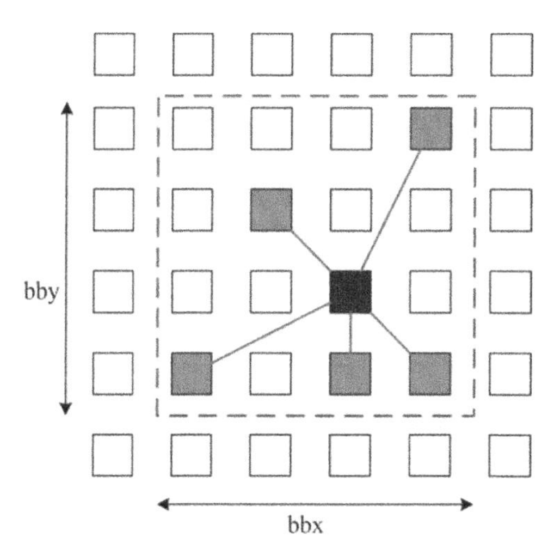

Figure 2.7: Bounding box of a hypothetical 6-terminal net [Betz et al., 1999]

widely used academic tool for FPGA technology mapping. FlowMap is able to find a depth-optimal solution in polynomial time. Later versions of FlowMap are further improved to optimize area and runtime of a boolean network while maintaining the same depth [J.Cong and Y.Ding, 1994b] [J.Cong and Y.Hwang, 1995] [J.Cong and Y.Ding, 2000]. The final output of FPGA technology mapping is a network of I/Os, LUTs and Flip-Flops.

Packing :

A mesh-based FPGA consists of an array of Configurable Logic Blocks (CLBs). Each CLB consists of a cluster of Basic Logic Elements (BLEs). A BLE consists of a Look-Up Table and a Flip-Flop. The packing phase, also called as clustering phase groups a Look-Up Table and a Flip-Flop in a BLE, and groups different BLEs into a cluster of BLEs. These BLEs or clusters of BLEs can then be directly mapped on the CLBs of an FPGA. The main optimization goal is to cluster the Look-Up Tables, Flip-Flops and BLEs in such a way that inter-cluster communication is minimized. Less inter-cluster communication ensures less routing resource utilization in an FPGA. The final output of packing is a network of I/Os and CLBs.

Placement :

The placement algorithm determines the position of CLB and I/O instances in a packed netlist on the respective CLB and I/O blocks on the FPGA architecture. The main goal of placement algorithm is to place connected blocks near each other so that minimum routing resources are required to route their connections. The placement algorithm can also serve to fulfill other architectural or optimization requirements, such as balancing the wire density across FPGA.

Three major types of commonly used placement algorithms include (i) min-cut (partitioning) based placement algorithm [A.Dunlop and B.Kernighan, 1985] [D.Huang and A.Kahng,

1997] (ii) analytical placement algorithm [G.Sigl et al., 1991] [C.J.Alpert et al., 1997], and (iii) simulated annealing based placement algorithm [S.Kirkpatrick et al., 1983] [C.Sechen and A.Sangiovanni-Vincentelli, 1985]. (i) The partitioning based placement approach is generally suitable for hierarchical FPGA architectures. The partitioner is recursively applied to distribute netlist instances between clusters. The aim is to reduce external communication and merge highly connected instances in the same cluster. (ii) Analytical placement algorithms commonly utilize a quadratic wire length objective function. Although, a quadratic objective is only an indirect measure of the wire length; its main advantage is that it can be minimized very efficiently and is thus suitable for handling large problems. A quadratic function does not give the best possible wire length; it is often followed by some local iterative improvement techniques. (iii) The simulated annealing placement algorithm uses the annealing concept for molten metal which is cooled down gradually to produce high quality metal objects. The simulated annealing algorithm is very effective at finding an acceptably good solution in a limited amount of time. This work concentrates on simulated annealing based placement algorithms.

The simulated annealing placement algorithm is good at approximating an acceptable placement solution for a netlist to be placed on an FPGA. A wire length cost function is used to measure the quality of placement. Netlist instances are initially placed randomly on the FPGA. Different instance moves are made to gradually improve the quality of placement. The "temperature" parameter of the algorithm cools down (decreases) systematically. At each temperature step, move operations are performed "Iteration Count" number of times. "Iteration Count" is proportional to the number of instances in a netlist. The main objective function of placer is to achieve a placement having minimum sum of half-perimeters of the bounding boxes of all the nets. Figure 2.7 shows a bounding box of a hypothetical 6-terminal net. An instance is randomly moved from one position to another; the change in the cost function is computed incrementally. If the cost decreases (improves), the move is always accepted. If the cost increases, the move can still be accepted. The probability of accepting a move that increases the cost is high during the initial phase of the algorithm. But this probability decreases gradually, until in the final phase only those moves are accepted which decrease the cost.

Routing :

Once the instances of a netlist are placed on FPGA, connections between different instances are routed using the available routing resources. The FPGA routing problem consists of routing the signals (or nets) in such a way that no more than one signal use the same routing resource. PathFinder [L.McMurchie and C.Ebeling, 1995] routing algorithm is commonly used for FPGAs. In order to perform routing on an FPGA architecture, the routing architecture is initially modeled as a directed graph where different nodes are connected through edges. Each routing wire of the architecture is represented by a node, and connection between two wires is represented by an edge. Figure 2.8 represents a small portion of routing architecture in the form of a directed graph. When a netlist is routed on the FPGA routing graph, each net (i.e connection of a driver instance with its receiver instances) is routed using a congestion driven Dijkstra's "Shortest Path" algorithm [T.Cormen et al., 1990]. Once all nets in a netlist

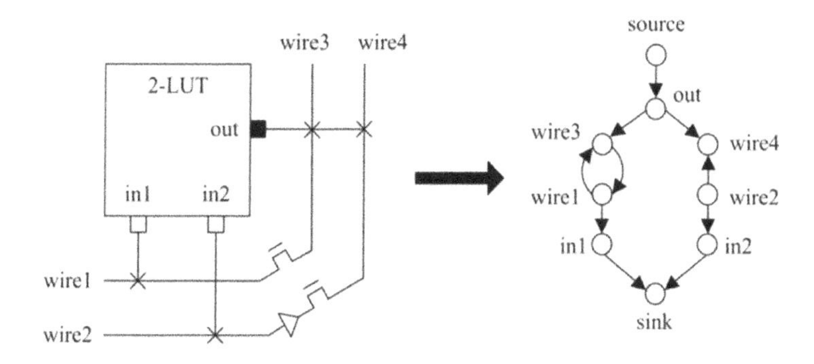

Figure 2.8: Modeling FPGA architecture as a directed graph [Betz et al., 1999]

are routed, one routing iteration is said to be completed. At the end of an iteration, there can be conflicts between different nets sharing the same nodes. The congestion parameters of the nodes are updated, and routing iterations are repeated until routing converges to a feasible solution (i.e all conflicts are resolved) or routing fails (i.e. maximum iteration count has reached, and few routing conflicts remain unresolved).

Bitstream generation :

Once a netlist is placed and routed on an FPGA, bitstream information is generated for the netlist. This bitstream is programmed on the FPGA using a bitstream loader. The bitstream of a netlist contains information as to which SRAM bit of an FPGA be programmed to 0 or to 1. The bitstream generator reads the technology mapping, packing and placement information to program the SRAM bits of Look-Up Tables. The routing information of a netlist is used to correctly program the SRAM bits of connection boxes and switch boxes.

2.2 Research trends in FPGAs

FPGA-based products are very effective for low to medium volume production, they are easy to program and debug, and have less NRE cost and less time-to-market. All these major advantages of an FPGA come through their reconfigurability. However the very same reconfigurability is the major cause of its disadvantages. The flexibility of FPGAs is mainly due to its reprogrammable routing network which takes 80 to 90% of the entire FPGA area [Betz et al., 1999]. The logic area is only 10 to 20% of the FPGA. Due to this reason FPGAs are much larger, slower and more power consuming than ASICs [I.Kuon and J.Rose, 2007]; thus they are unsuitable for high volume production, high performance or low power consumption.

Reconfigurable hardwares and FPGA architectures have many active research domains. A major aspect of research in reconfigurable hardwares revolves around decreasing the drawbacks of FPGAs, with or without compromising upon its major benefits. Following are few of the major tradeoff solutions that have been proposed in solving the area, speed, power and/or volume production problems of FPGAs.

- **Hard-Blocks:** Logic density of an FPGA is improved by incorporating dedicated hard-blocks in an FPGA. Hard-Blocks, or in other words small ASICs, in FPGAs increase their speed and reduce their overall area and power consumption. Hard-blocks can include multipliers, adders, memories, floating-point units etc. In this regard, [Beauchamp et al., 2006] have introduced embedded floating-point units in FPGAs, [C.H.Ho et al., 2006] have developed virtual embedded block methodology to model arbitrary embedded blocks on existing commercial FPGAs. [Hartenstein, 2001] has presented a brief survey of a decade of R&D on coarse grain reconfigurable hardware and their related compilation techniques. Figure 2.10 shows a commercial FPGA architecture that uses embedded hard-blocks.

- **Application Specific FPGAs:** The type of logic blocks and the routing network in an FPGA can be optimized to gain area and performance advantages for a given application domain (controlpath-oriented applications, datapath-oriented applications, etc). These types of FPGAs may include different variety of desired hard-blocks, appropriate amount of flexibility required for the given application domain or bus-based interconnects rather than bit-based interconnects. [Marshall et al., 1999] have presented a reconfigurable arithmetic array for multimedia applications, [Verma and Akoglu, 2007] have presented a coarse grained reconfigurable architecture for variable block size motion estimation and, [Ye and Rose, 2006] have used bus-based connections to improve density of FPGAs for datapath circuits. Figure 2.14 shows a reconfigurable arithmetic array for multimedia applications.

- **FPGA to Structured-ASIC:** The ease of designing and prototyping with FPGAs can be exploited to quickly design a hardware application on an FPGA. Later, improvements in area, speed, power and volume production can be achieved by migrating the application design from FPGA to other technologies such as Structured-ASICs. In this regard, Alter provides a facility to migrate its Stratix IV based application design to HardCopy IV [HardCopy, IV]. The eASIC Nextreme [eASIC, 2010] uses an FPGA-like design flow to map an application design on SRAM programmable LUTs, which are later interconnected through mask programming of few upper routing layers. cASIC [Compton and Hauck, 2007] explores the design space between ASIC and FPGA; configurable ASIC cores are designed to execute a given set of application designs at exclusive times. Tierlogic [TIERLOGIC, 2010] is a recently launched FPGA vendor that offers 3D SRAM-based TierFPGA devices for prototyping and early production. The same design solution can be frozen to a TierASIC device with one low-NRE custom mask for error-free transition to an ASIC implementation. The SRAM layer is placed on an upper 3D layer of TierFPGA. Once the TierFPGA design is frozen, the bitstream information is used to create a single custom mask metal layer that will replace the SRAM programming layer.

- **FPGA with processors:** A considerable amount of FPGA area can be saved by implementing the control path portion of a circuit on a microprocessor, and only the compute intensive datapath portion of a circuit is implemented on FPGAs. An FPGA is connected to a microprocessor in different ways (i) A soft processor is implemented on

FPGA reconfigurable resources (like shown in [NIOS, II], [MicroBlaze, 2010] and [Peter Yiannacouras and Rose, 2007]), (ii) a processor is incorporated in an FPGA as a dedicated hard-block (like AVR Processor integrated in Atmel FPSLIC [ATMEL, 2010] or PowerPC processors embedded in Xilinx Virtex-4 [Xilinx, 2010]), or (iii) an FPGA is attached with the pipeline of a processor to execute customized hardware instructions (like [Callahan et al., 2000] and [Jones et al., 2005]). Figure 2.13 illustrates a VLIW processor that supports application-specific hardware instructions.

- **Time-multiplexed signals:** Instead of using a dedicated routing track for routing a single signal, a routing wire is time-multiplexed and used by different signals at different times [Kapre et al., 2006] [Essen et al., 2009]. In this way, considerable amount of routing resources can be reduced to achieve area gains. Time multiplexing is handled by adding special hardware circuitry. These extra resources make time-multiplexing less attractive for commercial FPGA architectures where generally single-bit routing wires are used. However, these extra resources can be amortized across word-wide routing resources in coarse-grained reconfigurable arrays.

- **Time-multiplexed FPGAs:** The capacity or logic density of FPGAs is increased by executing different portions of a circuit on an FPGA in a time multiplexing mode [Trimberger et al., 1997] [Miyamoto and Ohmi, 2008]. An application design is divided into different sub-circuits, and each sub-circuit runs as a individual context of FPGA. The state information of each sub-circuit is saved in context registers before a new context runs on FPGA. Tabula [Tabula, 2010] is a recently launched FPGA vendor that provides time-multiplexed FPGAs.

The following section discusses different case-studies employing one of the above techniques.

2.2.1 Versatile Packing, Placement and Routing, VPR

Versatile Packing, Placement and Routing for FPGAs (commonly known as VPR) [V.Betz and J.Rose, 1997] [Betz et al., 1999] [A.Marquart et al., 1999] is the most widely used academic mesh-based FPGA exploration environment. It allows to explore mesh-based FPGA architectures by employing an empirical approach. Benchmark circuits are technology mapped, placed and routed on a desired FPGA architecture. Later, area and delay of FPGAs are measured to determine best architectural parameters. Different CAD tools in VPR are highly optimized to ensure high quality results; as poor CAD tools may lead to inaccurate architectural conclusions. Area and delay models are sufficiently accurate to compare the effect of different architectural changes.

The GILES project (Good Instant Layout of Erasable Semiconductors) [Padalia et al., 2003] [Kuon et al., 2005] generates the physical layout of an FPGA from the architecture specifications given as input to VPR. Developing a new FPGA is a challenging and time-consuming task. [Padalia et al., 2003] report that the creation of new FPGA requires 50 to 200 person years;

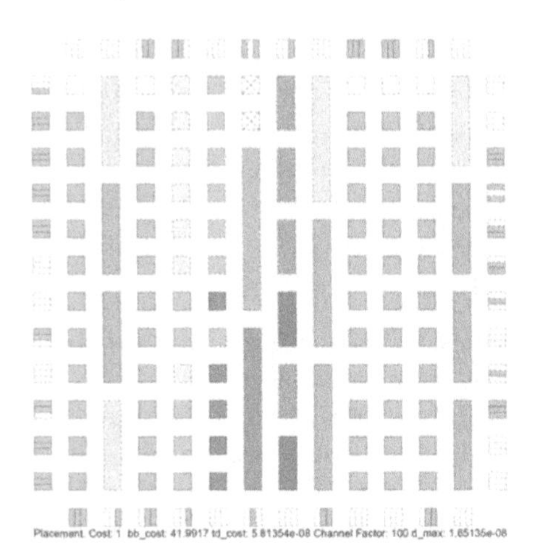

Figure 2.9: A Heterogeneous FPGA in VPR 5.0 [Luu et al., 2009]

thus increasing the overall time-to-market. It is thus an interesting option to significantly reduce the time-to-market of FPGAs at the expense of limited area penalty. GILES automatically generates a transistor-level schematic of an FPGA tile from a high-level architectural specification. The output of GILES is a mask-level layout of a single tile that can be replicated to form an FPGA array.

The latest version of VPR known as VPR 5.0 [Luu et al., 2009] supports hard-blocks (such as multiplier and memory blocks) and single-driver routing wires. Hard-blocks are restricted to be in one grid width column, and that column can be composed of same type of blocks. The height of a block must be an integral number of grid units. In case a block height is indivisible with the height of FPGA, some grid locations are left empty. Figure 2.9 illustrates a heterogeneous FPGA with 8 different kinds of blocks. VPR 5.0 also provides optimized electrical models for a wide range of architectures for different process technologies.

2.2.2 Madeo, a framework for exploring reconfigurable architectures

Madeo [Lagadec, 2000] is a design suite for the exploration of reconfigurable architectures. It includes a modeling environment that supports multi-grained, heterogeneous architectures with irregular topologies. Madeo framework initially allows to model an FPGA architecture. The architecture characteristics are represented as a common abstract model. Once the architecture is defined, the CAD tools of Madeo can be used to map a target netlist on the architecture. Madeo embeds placement and routing algorithms (the same as used by VPR [V.Betz and J.Rose, 1997]), a bitstream generator, a netlist simulator, and a physical layout generator. Madeo supports architectural prospection and very fast FPGA prototyping. Several FPGAs, including some commercial architectures (such as Xilinx Virtex family) and prospective ones (such as STMicro LPPGA) have been modeled using Madeo. The physical layout is produced as VHDL description.

2.2.3 Altera Architecture

Altera's Stratix IV [Stratix, IV] is a mesh-based FPGA architecture family fabricated in 40-nm process technology. Figure 2.10 shows the global architectural layout of Statix IV. The logic structure consists of LABs (Logic Array Blocks), memory blocks and digital signal processing (DSP) blocks. LABS are distributed symmetrically in rows and columns and are used to implement general purpose logic. The DSP blocks are used to implement full-precision multipliers of different granularities. The memory blocks and DSP blocks are placed in columns at equal distance with one another. Input and Output (I/Os) are located along the periphery of the device.

Logic array blocks (LABs) and adaptive logic modules (ALMs) are the basic building blocks of the Stratix VI device. They can be used to configure logic functions, arithmetic functions, and register functions. Each LAB consists of ten ALMs, carry chains, arithmetic chains, LAB control signals, local interconnect, and register chain connection lines. The internal LAB structure is shown in Figure 2.11. The local interconnect connects the ALMs belonging to the same LAB. The direct link allows a LAB to drive into the local interconnect of its left or right neighboring LAB. The register chain connects the output of ALM register to the adjacent ALM register in the LAB. A memory LAB (MLAB) is a derivative of LAB which can be either used just like a simple LAB, or as a static random access memory (SRAM). Each ALM in an MLAB can be configured as a 64x1, or 32x2 blocks, resulting in a configuration of 64x10 or 32x20 simple dual-port SRAM block. MLAB and LAB blocks always coexist as pairs in Stratix IV families.

The DSP blocks in Stratix IV are optimized to support signal processing applications such as Finite Impulse Response (FIR), Infinite Impulse Response (IIR), Fast Fourier Transform functions (FFT) and encoders etc. Stratix IV device has two to seven columns of DSP blocks that can implement multiplication, multiply-add, multiply-accumulate (MAC) and dynamic arithmetic or logical shift functions. The DSP block supports 9x9, 12x12, 18x18 and 36x36 multiplication operations. The Statix IV devices contain three different sizes of embedded SRAMs. The memory sizes include 640-bit memory logic array blocks (MLABs), 9-Kbit M9K blocks, and 144-Kbit M144K blocks. The MLABs have been optimized to implement filter delay lines, small FIFO buffers, and shift registers. M9K blocks can be used for general purpose memory applications, and M144K are generally meant to store code for a processor, packet buffering or video frame buffering.

2.2.4 Altera HardCopy

Altera gives provision to migrate FPGA-based applications to Structured-ASIC. Their Structured-ASIC is called as HardCopy [HardCopy, IV]. The main theme is to design, test and even initially ship a design using an FPGA. Later, the application circuit that is mapped on the FPGA can be seamlessly migrated to HardCopy for high volume production. Their latest HardCopy-IV devices offer pin-to-pin compatibility with the Stratix IV prototype, making

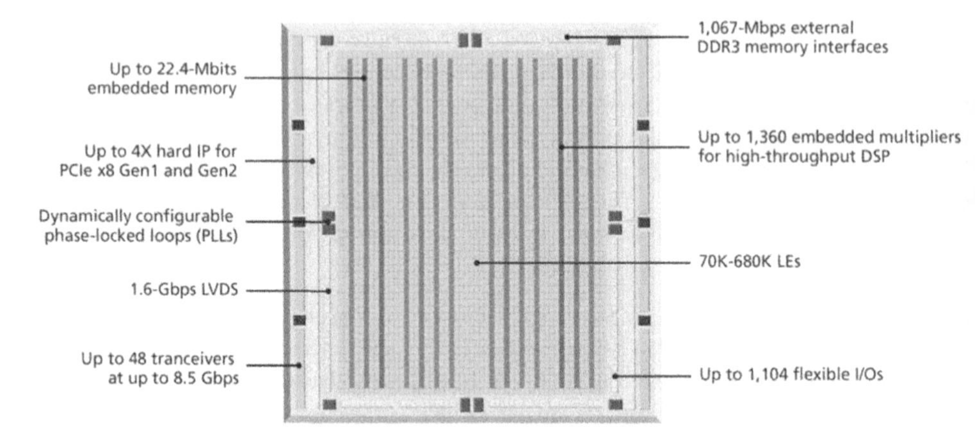

Figure 2.10: Stratix IV architectural elements

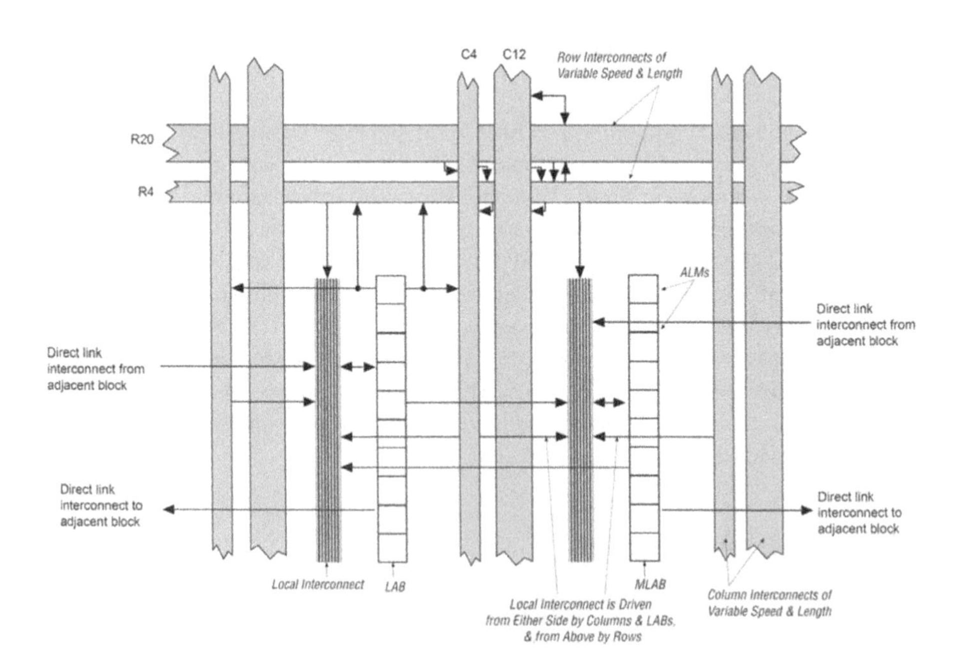

Figure 2.11: Stratix IV LAB Structure

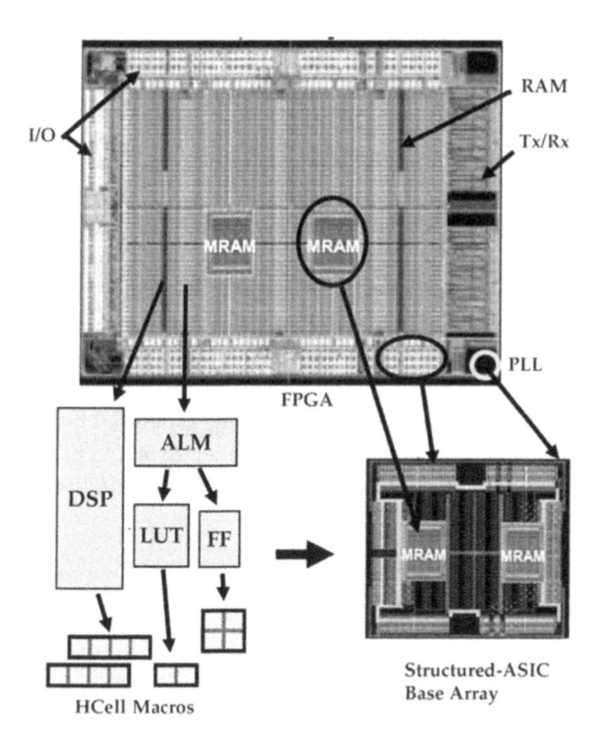

Figure 2.12: FPGA/Structured-ASIC (HardCopy) Correspondence [Hutton et al., 2006]

them drop-in replacements for the FPGAs. Thus, the same system board and softwares developed for prototyping and field trials can be retained, enabling the lowest risk and fastest time-to-market for high-volume production. Moreover, when an application circuit is migrated from Stratix IV FPGA prototype to Hardcopy-VI, the core logic performance doubles and power consumption reduces by 50%.

The basic logic elements in an SRAM-based FPGA comprise of LUTs and Flip-Flops. The logic functionality is implemented on these LUTs which are optionally registered. On the other hand, the basic logic unit of HardCopy is termed as HCell. It is similar to FPGA logic cell (LAB) in the sense that the fabric consists of a regular pattern which is formed by tiling one or more basic cells in a two dimensional array. The difference is that HCell has no configuration overhead. Different HCell candidates can be used, ranging from fine-grained NAND gates to multiplexors and coarse-grained LUTs. An array of such HCells candidates, and a general purpose routing network which interconnects them is laid down on the lower layers of the chip. Specific layers are then reserved to form via connections or metal lines which are used to customize the generic array into specific functionality. Figure 2.12 illustrates the correspondence between an FPGA floorplan and a compatible structured ASIC base array. There is a one to one layout-level correspondence between MRAMs, phase-lock loops (PLLs), embedded memories, transceivers, and I/O blocks. The soft-logic DSP multipliers and logic cell fabric of the FPGA are re-synthesized to structured ASIC fabric (HCells). However, they remain functionally and electrically equivalent in FPGAs and HardCopy ASICs.

2.2.5 Configurable ASIC Cores (cASIC)

Configurable ASIC Core (cASIC) [Compton and Hauck, 2007] is a reconfigurable device that can implement a limited set of circuits which operate at mutually exclusive times. cASICs are intended as accelerator in domain-specific systems-on-a-chip, and are not designed to replace the entire ASIC-only chip. The host would execute software code, whereas compute-intensive sections can be offloaded to one or more cASICs. For that reason, cASICs implement only data-path circuits and thus supports full-word blocks only (such as 16-bit wide multipliers, adders, RAMS, etc). Since the set of circuits supported by a cASIC are limited, cASICs are significantly smaller than an FPGA implementation. As hardware resources are shared between different netlists, cASICs are even smaller than the sum of the standard-cell based ASIC areas of individual circuits.

Automatic generation of cASIC cores occurs in two phases. The logic phase determines the computation needs of the given set of application netlists. Different computational components are generated which may include ALUs, RAMs, multipliers, registers, etc. These logic resources are shared by all the application netlists. The logic components are properly ordered along the one-dimensional datapath so that minimum routing resources are required. The routing phase then creates wires and multiplexors to connect logic and I/O components. The objective of routing phase is to minimize area by sharing wires between different netlists while adding as few multiplexors/demultiplexors as possible. Different heuristic algorithms are used to maximize wire sharing. Experiments show that configurable ASIC hardware is on average 12.3x smaller than an FPGA solution with embedded multiplier, and 2.2x smaller than a standard cell implementation of individual circuits.

2.2.6 FPGA based processors

Considerable amount of FPGA area can be reduced by incorporating a microprocessor in an FPGA. A microprocessor can execute any less compute intensive task, whereas compute-intensive tasks can be executed on an FPGA. Similarly, a microprocessor based application can have huge speed-up gains if an FPGA is attached with it. An FPGA attached with a microprocessor can execute any compute intensive functionality as a customized hardware instruction. These advantages have compelled commercial FPGA vendors to provide microprocessor in their FPGAs so that complete system can be programmed on a single chip. Few vendors have integrated fixed hard processor on their FPGA (like AVR Processor integrated in Atmel FPSLIC [ATMEL, 2010] or PowerPC processors embedded in Xilinx Virtex-4 [Xilinx, 2010]). Others provide soft processor cores which are highly optimized to be mapped on the programmable resources of FPGA. Altera's Nios [NIOS, II] and Xilinx's Microblaze [MicroBlaze, 2010] are soft processor meant for FPGA designs which allow custom hardware instructions. [Peter Yiannacouras and Rose, 2007] have shown that considerable area gains

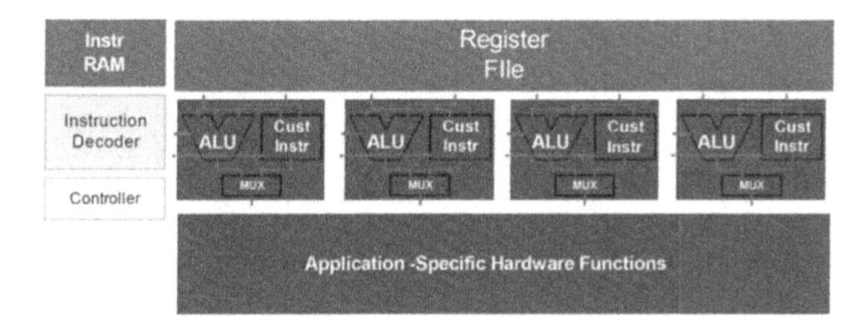

Figure 2.13: A VLIW Processor Architecture with Application Specific Hardware Functions [Jones et al., 2005]

can be achieved if these soft processors for FPGAs are optimized for particular applications. They have shown that unused instructions in a soft processor can be removed and different architectural tradeoffs can be selected to achieve on average 25% area gain for soft processors required for specific applications.

Reconfigurable units can also be attached with microprocessors to achieve execution time speedup in software programs. There is a famous 90/10 rule, which states that 90% of the program's execution time is spent in 10% of the code. So the major aim is to convert the 10% code into hardware logic and implement it as a hardware function to achieve very large gains. Whereas the remaining 90% of the code runs on a microprocessor. [Callahan et al., 2000], [Sima et al., 2001] and [Jones et al., 2005] have incorporated a reconfigurable unit with microprocessors to achieve execution-time speedup.

[Jones et al., 2005] have proposed a VLIW processor which can support application specific customized instructions through a reconfigurable hardware block. VLIW processors have a single instruction controller that dispatches different operations to several functional units that share a single register file. All these functional units execute in parallel. Ideally, many instructions can execute in parallel. But the application must also exhibit high Instruction Level Parallelism (ILP), so that control and data dependencies do not limit the performance improvements. The proposed architecture, illustrated in Figure 2.13, is a 4 way VLIW processor (4 functional units), with hardware resources meant for implementing application specific hardware functions. All the four functional blocks and the hardware functions are linked together through the register file. The hardware block is able to read 16 operands from any of the 32 registers and write back 8 results into any of the registers.

A compilation process is developed for the VLIW processor with hardware functions. The C code to be implemented is profiled to find the computational intensive kernels. Behavioral or high-level synthesis is applied to automatically transform them into combinatorial logic. The remaining code is transformed by the VLIW compiler and assembler, and implemented on the VLIW processor. The VLIW processor and the hardware block are implemented on Altera Stratix II FPGA due to its support for high-speed DSP blocks. The VLIW processor with hardware functions have shown a maximum speedup of 230 times, and an average speedup of 63 times for computational kernels from a set of DSP benchmarks. The overall maximum

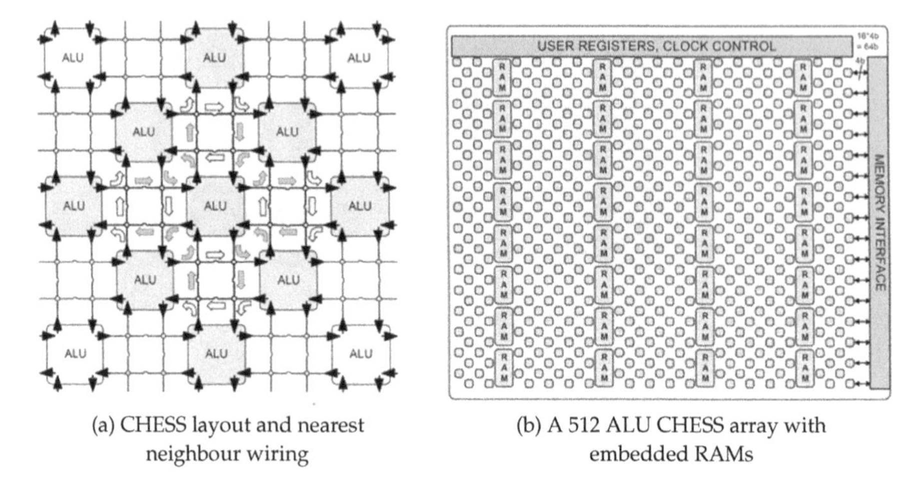

(a) CHESS layout and nearest neighbour wiring

(b) A 512 ALU CHESS array with embedded RAMs

Figure 2.14: A Reconfigurable Arithmetic Array for Multimedia Applications

speedup of 30 times and an average speedup of 12 times is shown for DSP benchmarks from MediaBench.

2.2.7 A Reconfigurable Arithmetic Array for Multimedia Applications, CHESS

[Marshall et al., 1999] have proposed a reconfigurable arithmetic array for multimedia applications. Due to its layout, this architecture is also termed as CHESS. The principal goal of CHESS was to increase arithmetic computational density, to enhance the flexibility, and to increase the bandwidth and capacity of internal memories significantly beyond the capabilities of existing commercial FPGAs. These goals were achieved by proposing an array of ALUs with embedded RAMs as shown in Figure 2.14. Each ALU is 4-bit wide and supports 16 instructions. No run time configuration is required, as ALU instruction can be changed dynamically. The CHESS layout of ALUs ensures strong local connection with 8 adjacent ALUs, as shown in Figure 2.14(a). A switch box lies between adjacent ALUs. The switch box contains 64 connections, which can also act as 16W*4 RAMs. Dedicated RAM blocks are placed in columns at equal distance with each other, as shown in Figure 2.14(b). The major advantage of CHESS is that routing network takes only 50% of area.

2.2.8 Reconfigurable Pipelined Data paths, Rapid

Reconfigurable pipelined data paths (known as Rapid) [Ebeling et al., 1996] is a coarse-grained, field programmable architecture for constructing deep computational pipelines. Rapid architecture can efficiently implement applications related to media, signal processing, scientific computing and communications. Rapid consists of a linear array of functional units that are interconnected through a programmable segmented bus network. These

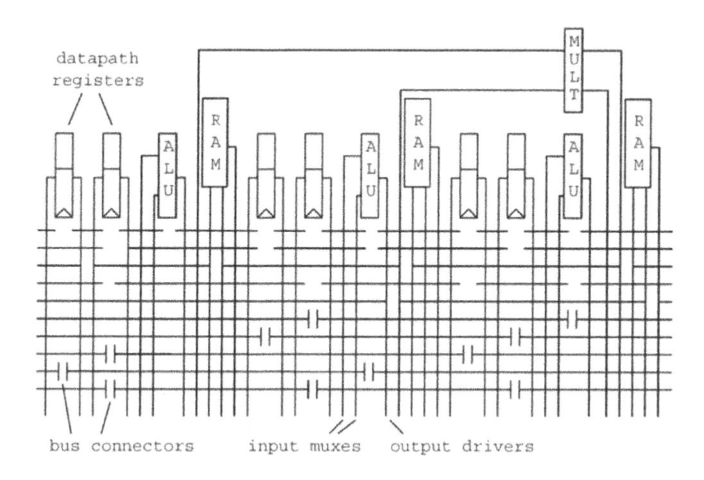

Figure 2.15: A basic cell of RaPiD [Ebeling et al., 1996]

coarse-grained functional units and the bus interconnect are used to implement a data path circuit. The rapid datapath is usually divided into identical units, called as cells, that are replicated to form a complete datapath. A typical rapid cell is shown in Figure 2.15. Each cell can contain hundreds of functional units ranging in complexity from simple general purpose register to multi-output booth-encoded multipliers. All functional units are linearly arranged and are connected to a segmented bus network that runs the entire length of datapath. The functional unit outputs are connected to optional delay units which can be configured to 0 to 3 register delays. This feature allows creation of deep pipelines. The input multiplexors on the segmented bus network are used to give input to functional unit. The Rapid architecture is configured through programming bits that are divided into soft bits and hard bits. The hard bits are the SRAM bits which do not change during the execution of a single application. The soft bits can change after each clock cycle. A pipelined control path generate these soft bits to control the datapath circuit. The control logic achieved through static and dynamic bits substantially reduces the control overhead as compared to FPGA-based and general-purpose processor architectures. Rapid architecture is programmed through Rapid-C [Ebeling, 2002], a C-like language with extensions to specify parallelism, partitioning and data movement. Rapid-C programs may contain several nested loops. Outer loops are transformed into sequential code for address generators, inner loops are mapped on Rapid cells.

2.2.9 Time-Multiplexed Signals

Time-Multiplexed signals can schedule different signals on the same routing wires. [Essen et al., 2009] have analyzed the design tradeoffs involved in static vs time-multiplexed routing for coarse-grained reconfigurable arrays. Unlike commercial FPGA architectures, where routing resources are configured in exactly one way for the entire run of a single application, time-multiplexing allows routing configuration to be changed on a cycle by cycle basis. Iterating through the schedule of configurations, a scheduled channel changes its communication pattern on each cycle. Time-multiplexing of scheduled signals requires additional configuration memories and some control circuitry. These extra resources make time-multiplexing less attractive for commercial FPGA architectures where generally single-bit routing wires are used. However, when applied to coarse-grained reconfigurable arrays, these extra resources can be amortized across word-wide resources.

The optimal tradeoff between scheduled and static resources depends on the word-width of the interconnect, since the overheads associated with some techniques may be much larger in a 1-bit interconnect than with a 32-bit interconnect. To explore this possibility, the effect of different word-widths is measured on the area and power consumption of the time-multiplexed coarse-grained architectures. It is found that for 32-bit word-wide interconnects, going from 100% statically configured to 100% scheduled (time-multiplexed) channels reduce the channel width to 0.38x the baseline. This in turn reduces the energy to 0.75x, the area to 0.42x, and the area-energy product to 0.32x, despite the additional configuration overhead. This is primarily due to amortizing the overhead of a scheduled channel across a multi-bit signal. It is important to note that as the datapath width is reduced, approaching the single bit granularity of an FPGA, the scheduled channel overhead becomes more costly. It is found that for datapath widths of 24-, 16-, and 8-bit, converting from fully static to fully scheduled reduces area-energy product to 0.34x, 0.36x, and 0.45x, respectively.

Another factor that significantly affects the best ratio of scheduled versus static channels is the maximum degree of time-multiplexing supported by the hardware, i.e. its maximum Initiation Interval (II). Supporting larger II translates into more area and energy overhead for scheduled channels. It is shown that for a 32-bit datapath, supporting an II of 128 is only 1.49x more expensive in area-energy than an II of 16; a fully scheduled interconnect is still a good choice. However, for an 8-bit datapath and a maximum II of 128, 70% static (30% scheduled) achieves the best area-energy performance, and fully static is better than fully scheduled.

2.2.10 Time-Multiplexed FPGA

Time-multiplexed FPGAs increase the capacity of FPGAs by executing different portions of a circuit in a time-multiplexed mode. [Trimberger et al., 1997] have proposed a time-multiplexed FPGA architecture. A large circuit is divided into sub-circuits; each sub-circuit is sequentially executed on a time-multiplexed FPGA. Each sub-circuit runs as a separate context on the FPGA. Such an FPGA stores a set of configuration bits for all contexts. A context is shifted simply by using the SRAM bits dedicated to a particular context. The combinatorial and sequential outputs of a sub-circuit that are required by other sub-circuits are saved in context registers which can be easily accessed by sub-circuits at different times.

Time-Multiplexed FPGAs increase their capacity by actually adding more SRAM bits rather than more CLBs. These FPGAs increase the logic capacity by dynamically reusing the hard-

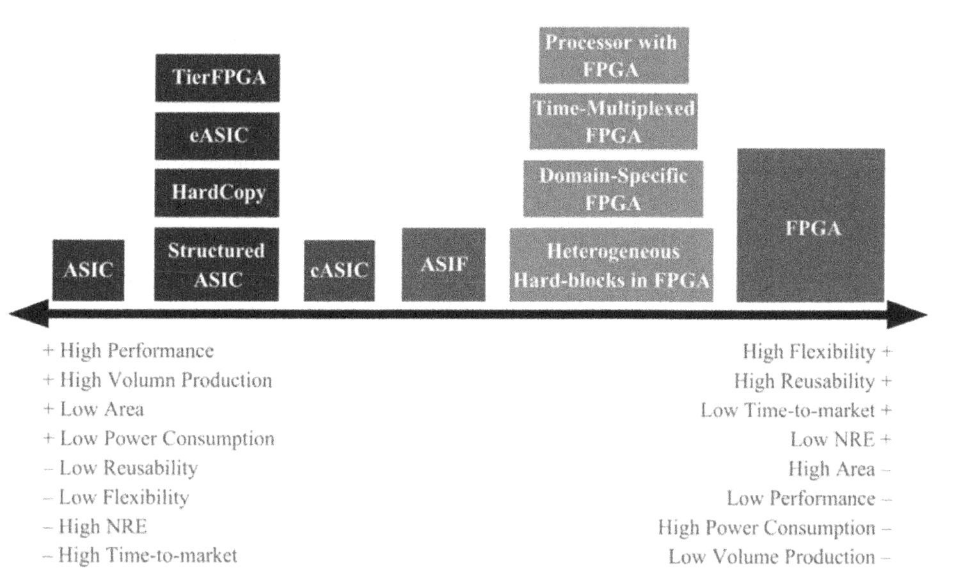

Figure 2.16: Comparison of different solutions used to reduce ASIC and FPGA drawbacks

ware. The configuration bits of only the currently executing context are active, the configuration bits for the remaining supported contexts are inactive. Intermediate results are saved and then shared with the contexts still to be run. Each context takes a micro-cycle time to execute one context. The sum of the micro-cycles of all the contexts makes one user-cycle. The entire time-multiplexed FPGA or its smaller portion can be configured to (i) execute a single design, where each context runs a sub-design, (ii) execute multiple designs in time-multiplexed modes, or (iii) execute statically only one single design.

Tabula [Tabula, 2010] is a recently launched FPGA vendor that provides time-multiplexed FPGAs. It dynamically reconfigures logic, memory, and interconnect at multi-GHz rates with a Spacetime compiler.

2.3 Conclusion

This chapter has initially presented a brief introduction of a traditional FPGA architecture, and related software flow to program hardware designs on the FPGA. It has also described various approaches that have been employed to reduce few disadvantages of FPGAs and ASICs, with or without compromising upon their major benefits. Figure 2.16 presents a rough comparison of different solutions used to reduce the drawbacks of FPGAs and ASICs. The next few chapters of this book will focus on the exploration of FPGA architectures using hard-blocks, application specific Inflexible FPGAs (ASIF), and their automatic layout generation methods.

This work presents a new environment for the exploration of heterogeneous hard-blocks in an FPGA. Hard-blocks are used in commercial FPGA architectures to reduce area, power and performance gaps between FPGAs and ASICs. Specialized hard-blocks, their architectural floor-planning, and specialized routing network can also be used to design domain specific FPGA architectures [Ebeling et al., 1996] [Marshall et al., 1999]. Unlike existing exploration environments [Luu et al., 2009] [Lagadec, 2000], the heterogeneous exploration environment proposed in this work can perform automatic optimization of architecture floor-planning for a given set of application circuits.

Altera [Altera, 2010] has proposed a new idea to prototype, test, and even ship initial few designs on an FPGA, later the FPGA based design can be migrated to Structured-ASIC (known as HardCopy) for high volume production. Other commercial vendors such as eASIC [eASIC, 2010] and TierLogic [TIERLOGIC, 2010] also propose a similar solution. However, migration of an FPGA-based product to Structured-ASIC supports only a single application design. HardCopy, eASIC and TierFPGA totally lose the quality of an FPGA to use the same hardware for executing multiple applications at different times. An ASIF retains this property, and can be a possible future extension for the migration of FPGA-based applications to Structured-ASIC. The concept of an ASIF is similar to a cASIC [Compton and Hauck, 2007], which can execute multiple applications at different times. However unlike the bottom-up insertion technique for the generation of cASIC, the top-down removal technique of an ASIF can be applied to any existing FPGA architecture. Thus when an FPGA-based product is in the final phase of its development cycle, and if the set of circuits to be mapped on the FPGA are known, the FPGA can be reduced to an ASIF for the given set of application designs.

This work also presents automatic layout generation techniques to reduce time-to-market and NRE costs of domain-specific FPGA and ASIF architectures.

3

Heterogeneous FPGA Exploration Environment

This chapter presents an environment for the exploration of 2D mesh-based heterogeneous FPGA architectures. An architecture description mechanism allows to define various architectural parameters including definition of new heterogeneous blocks, position of these blocks on the architecture and the choice of routing network. Once the initial architecture is defined, a software flow places and routes a target netlist on the generated architecture. The placement algorithm not only changes the position of netlist instances on their respective blocks on the architecture, but it also refines the position of blocks on the architecture. The position of blocks on the FPGA architecture, also called as floor-planning, can also be optimized for multiple netlists. A set of DSP test-benches are used to compare the effect of different floor-plannings on the area of mesh-based heterogeneous FPGA architectures.

3.1 Introduction and Previous Work

Embedded hard-blocks (HBs) are commonly used in FPGAs to decrease area, speed and power gap between ASICs and FPGAs [I.Kuon and J.Rose, 2007] [Underwood and Hemmert, 2004]. Commercial FPGA vendors, such as Xilinx [Xilinx, 2010] and Altera [Altera, 2010], use different variety of hard-blocks (HB) in their FPGAs. These vendors provide a range of FPGA device variants to fulfill varying domain-specific requirements of their customers. Floor-planning of these FPGAs is optimized by placing similar type of blocks in columns; such FPGAs are termed here as column-based. The widths of entire columns of a column-based FPGA can be appropriately adjusted to generate a compact layout. However, column-based floor-planning restricts each column to support only one type of HB. Due to this limitation, the number of hard-blocks in an FPGA must be in multiples of hard-blocks

H. Parvez and H. Mehrez, *Application-Specific Mesh-based Heterogeneous FPGA Architectures*,
DOI 10.1007/978-1-4419-7928-5_3, © Springer Science+Business Media, LLC 2011

that can fit in a column. Some hard-blocks might remain unused if customer requirement does not exactly match the number of hard-blocks on the architecture. Moreover, communication between different hard-blocks, placed in distinct columns, require additional routing resources; thus FPGA area increases. These area losses will become severe when more types of hard-blocks are required in a column-based FPGA to support new domain-specific requirements of customers. In this context, this work compares different floor-plannings of domain-specific heterogeneous FPGA architectures.

The Versatile Packing, Placement and Routing tool (VPR) [V.Betz and J.Rose, 1997] has been widely used for the exploration of fine-grained FPGA architectures. The initial version of VPR did not support heterogeneous blocks. However, [Beauchamp et al., 2006] and [Jamieson and J.Rose, 2006] have extended VPR to explore specific heterogeneous architectures. Similarly [C.H.Ho et al., 2006] have developed virtual embedded block methodology (VEB) to model arbitrary embedded blocks on existing commercial FPGAs. One of the key advantages of VEB is that new embedded blocks can be tested on commercial FPGAs. But the VEB methodology can only be used with existing commercial architectures. This deficiency has been resolved in [Yu, 2007] by incorporating VEB methodology in VPR; thus enabling support of architectures other than commercial FPGAs. [S.Dai and E.Bozorgzadeh, 2006] has also developed a CAD tool for FPGAs with Embedded Hard Cores. The recently released version of VPR [Luu et al., 2009] now supports heterogeneous hard-blocks such as multipliers and RAMS.

All this previous work propose a pre-determined floor-planning organization and does not consider the problem of optimizing the floor-planning automatically. Our major contribution consists of proposing an environment that refines architecture floor-planning according to netlist requirements. Although an FPGA floor-planning can be achieved manually, but such a task becomes more difficult when different variety of blocks are to be integrated in architecture. The manual floor-planning becomes further complicated when an FPGA architectures is required to be optimized for a given set of application netlists. The method proposed in this work places all the netlists simultaneously and changes the architecture floor-planning to get a trade-off architecture for given set of netlists. This technique of architectural floor-planning for multiple netlists is previously used to explore one dimensional, segmented-bus based configurable ASIC Cores [Compton and Hauck, 2007]. This work extends their methodology to explore two dimensional, mesh-based heterogeneous FPGA architectures. The Heterogeneous FPGA exploration environment presented in this work is extended from a fine-grained homogeneous FPGA exploration environment [Marrakchi, 2008].

3.2 Architecture Exploration Environment

A heterogeneous FPGA architecture is represented on a grid of equally sized SLOTS called as the slot-grid. BLOCKS of different sizes can be mapped on this grid as shown in Figure 3.1. A BLOCK can be a soft-block like a Configurable Logic Block (CLB), or a hard-block like an adder, multiplier or RAM etc. Each BLOCK occupies one or more SLOTS. A routing channel

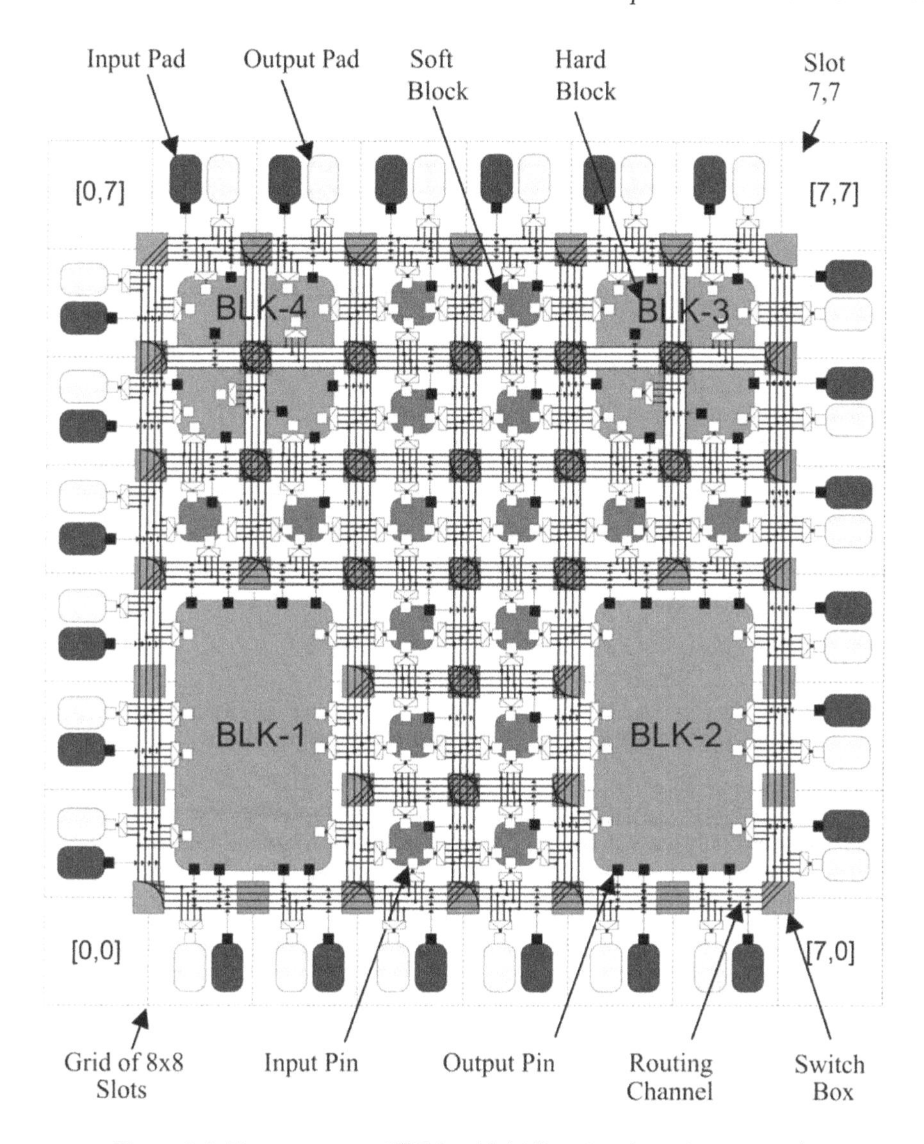

Figure 3.1: Heterogeneous FPGA with bidirectional routing network

crosses every two neighboring SLOTs. A BLOCK occupying more than one SLOT can allow or disallow routing channel to pass through it. BLK-1 and BLK-2 in Figure 3.1 do not allow routing channel to pass through them, whereas BLK-3 and BLK-4 allow routing channel to pass through them. Once the architecture is defined, architecture description parameters and a target netlist is passed to the software flow. The software flow maps netlist instances on BLOCKS of their respective types. The PLACER, a software module, refines the position of BLOCKS on the slot-grid and the placement of netlist instances on the BLOCKS. The

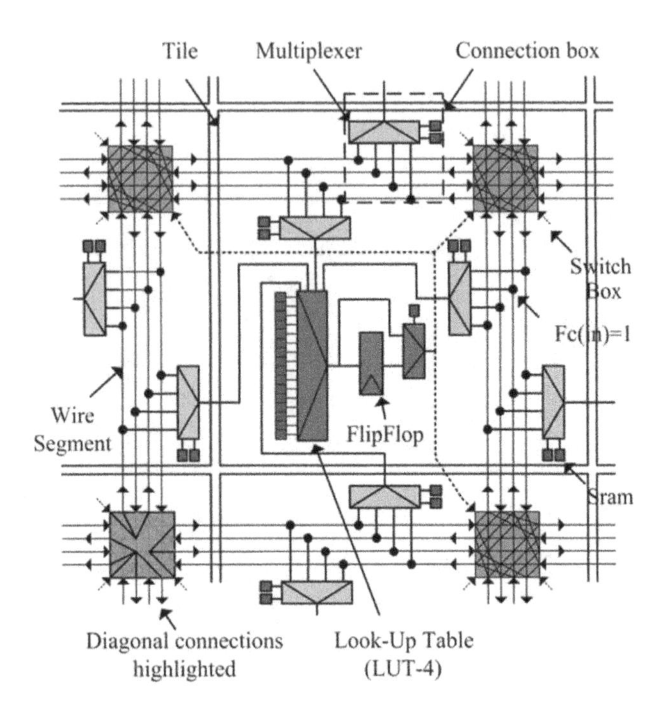

Figure 3.2: A single CLB block with unidirectional routing network

PLACER can optimize FPGA floor-planning for a given set of netlists that are to be mapped on the FPGA at mutually exclusive times. This is done by allowing multiple instances to map on a single BLOCK. However, instances belonging to the same netlist cannot be mapped on the same BLOCK. After placement and floor-planning, ROUTER routes the netlist on the architecture.

3.3 Architecture Description Mechanism

An architecture description mechanism is used to define FPGA architecture parameters. These parameters are grouped in an architecture description file. Few major architecture description parameters are shown in Table 3.1 and 3.2. The parameters Nx and Ny define the size of the slot-grid. Channel_Type is used to select a unidirectional mesh [Lemieux et al., 2004] or a bidirectional mesh [V.Betz and J.Rose, 1997] routing network. The channel width of the routing network is either set to a constant value (using the parameter Channel_Width), or a binary search algorithm searches a minimum possible channel width between minimum (Channel_Width_Min) and maximum (Channel_Width_Max) channel width limits. In case of unidirectional mesh, the channel width remains in multiples of 2. The position of BLOCKS can be set to an absolute position on the slot-grid (by using the parameter Set_Block). This parameter takes the name of the block and the position on the slot-grid

	Name	Description
1.	Nx num	Slots in the slot-grid in X direction (num >1)
2.	Ny num	Slots in the slot-grid in Y direction (num >1)
3.	Input_Rate	Number of Input pads in each slot on the periphery of slot-grid
4.	Output_Rate	Number of Output pads in each slot on the periphery of slot-grid
5.	Channel_Type T	T is unidirectional or bidirectional
6.	Binary_Search F	Binary search flag (F is true or false)
7.	Channel_Width num	Channel width if Binary_Search = false (num > 1)
8.	Channel_Width_Min num	Minimum channel width if Binary_Search = true (num > 1)
9.	Channel_Width_Max num	Maximum channel width if Binary_Search = true (num > 1)
10.	Set_Block blk X Y	Place a block named blk at a slot position (X,Y) of slot-grid
11.	Set_Block_Auto blk N	Place N instances of blk on first available position of slot-grid
12.	Fix_Block_Positions F	The Blocks are movable or fixed (F is true or false)
13.	Block_Jump F	If Blocks are moveable, blocks can be moved (F is true or false)
14.	Block_Rotate F	If Blocks are moveable, blocks can be rotated (F is true or false)
15.	Column_Move W s	If Blocks are moveable, a column can be moved (W is width of the column, s is the starting horizontal slot position of column)
16.	Define_Block blk	Block definition (See Table 3.2)

Table 3.1: Architecture Description File Parameters

where it should be placed. Another option to place BLOCKS on the slot-grid is by using the parameter Set_Block_Auto. This parameter automatically places N copies of a BLOCK on the first available position on the slot-grid. The BLOCKS on the slot-grid can be either fixed to an initial position or set as moveable (by using the parameter Fix_Block_Positions). In case the BLOCKS are moveable, the PLACER can refine their position on the slot-grid. The parameter Block_Jump allows the PLACER to move BLOCKS on the slot-grid. The parameter Block_Rotate allows the PLACER to rotate BLOCKS at their own axis. The parameter Column_Move allows to move a complete column from one position to another. Column_Move parameter requires the width of column, W (i.e number of slots as column width), and the starting horizontal slot position of the column, s. All the BLOCKS in a column must be within the boundaries of the column. This parameter can be repeated if multiple columns are required to be moved.

A new BLOCK can be defined in the architecture description file using the Define_block parameter. The BLOCK definition parameters are shown in Table 3.2. Each BLOCK is given a name, a size (number of slots occupied), a rotation flag and input/output pins. The rotation flag allows the rotation of individual BLOCK by the PLACER. This rotation flag permits to turn off the rotation of a particular type of block when the global rotation is turned on. Each pin of the BLOCK is given a name, a CLASS number, a direction and a slot position on the BLOCK to which this pin is connected. Figure 3.1 gives the pictorial view of different BLOCKS mapped on the slot-grid. The detailed block definition of two simple BLOCKS is shown in Figure 3.3. The BLOCK "CLB" in the figure is composed of only 1 slot. It has four input pins on the LEFT, TOP, RIGHT and BOTTOM side of the slot. The output pin is connected to both TOP and RIGHT sides. All the pins of CLB are connected to the only

Definition	Description
Define_Block block_name	
X_Slots X	X is number of slots occupied in X direction
Y_Slots Y	Y is number of slots occupied in Y direction
Rotate F	If rotation allowed. F is true or false
Area area	Area of the block
Sram num	Number of SRAMs used by the block
Pin_Input Name Class X Y Dir	X, Y gives slot position of block on which the pin
.....	is located.
Pin_Output Name Class X Y Dir	Dir is direction of pin on slot. The direction
.....	of a pin can be left, top, right or bottom
End_Define_Block	

Table 3.2: Block definition

available slot i.e (0,0), however they have different directions. All the input pins of CLB have same CLASS number. Pins having the same CLASS number are considered equivalent. A driver net targeting a receiver pin of a BLOCK can be routed to any of the receiver pins of the BLOCK having the same CLASS. The BLOCK "MUL4" in the figure is composed of 6 slots, having width equal to 2 slots and height equal to 3 slots. Four input pins of MUL4 are placed on the LEFT side of slots (0,0), (0,1) and (0,2); the other four input pins are placed on the RIGHT side of slots (1,0), (1,1) and (1,2). The output pins are placed on the BOTTOM side of slots (0,0) and (1,0), and on the TOP side of slots (0,2) and (1,2). The parameter "Area" assigns area to the defined block. This area is used by the area model to compute the total area of FPGA. The parameter "Sram" is used by the layout generator in Chapter 7. It defines the total SRAMs used by a block. A CLB with a LUT-4 and an optional Flip-Flop requires 17 SRAMs, whereas a MUL4 requires no SRAMs.

3.4 Software Flow

Once the architecture is properly defined, a software flow maps a target netlist on the FPGA architecture. Different modules of the software flow are shown in figure 3.4. The input to this software flow is either a 'C' code or a structural netlist in VST (structured VHDL) format. GAUT [GAUT, 2010] performs high level synthesis to transform C Code to RTL. GAUT has an extensive set of hard-blocks that are used in the generated VHDL code. The current version of GAUT is restricted to generate hard-blocks with uniform bit-widths for its input and output pins. Thus a 16-bit multiplier has two 16-bit inputs and one 16-bit output (instead of a 32-bit output). The VHDL generated through GAUT is synthesized to VST (structured VHDL) through Synopsys [Synopsys, 2010]. The netlist in VST format is composed of traditional standard cell library instances and hard-block instances. The VST2BLIF tool converts the VST file to BLIF format. Later, PARSER-1 removes all the instances of hard-blocks and passes the remaining netlist to SIS [E.M.Sentovich et al., 1992] for synthesis into 4 input Look-Up Table format. All the dependence between the hard-blocks and the remaining netlist is

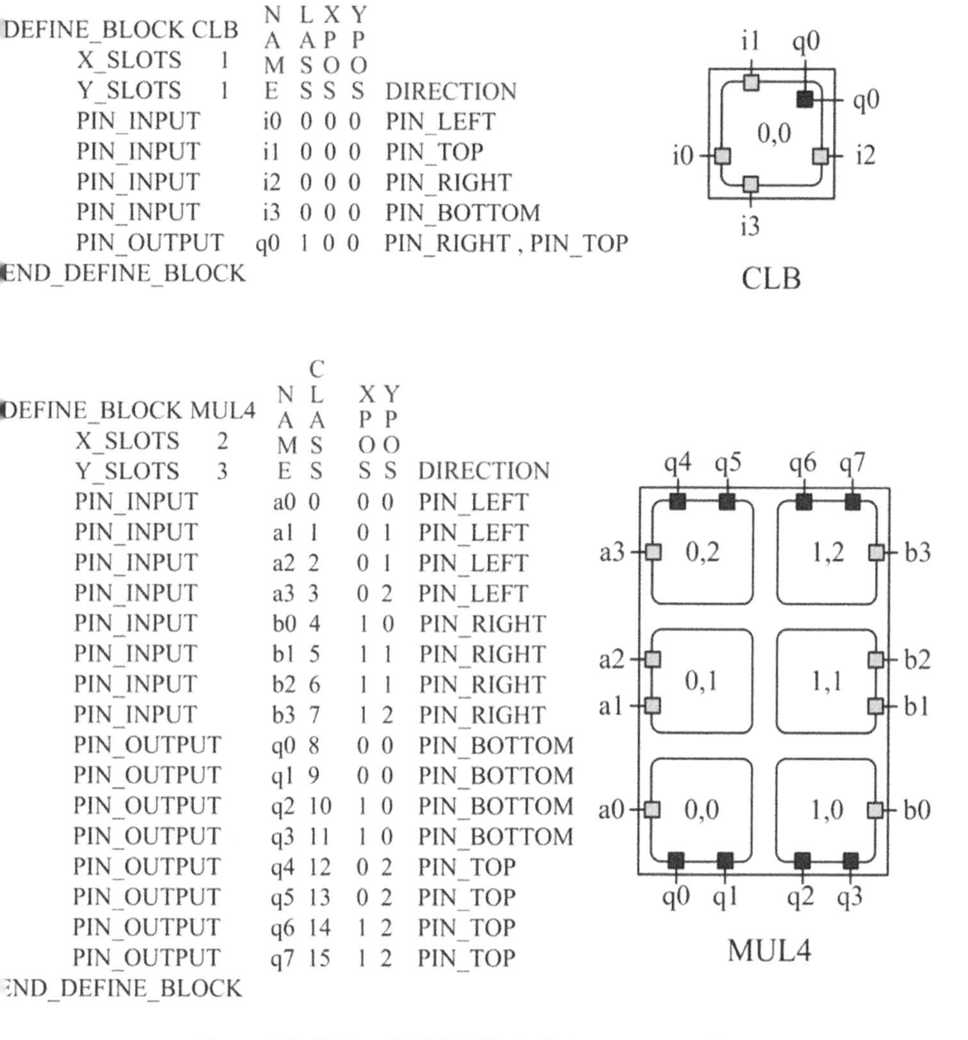

Figure 3.3: CLB and MUL4 Block definition example

preserved by adding new input and output pins to the main netlist. SIS generates a network of LUTs and Flip-Flops, which are later packed into CLBs through T-VPACK [V.Betz and J.Rose, 1997]. T-VPACK generates a netlist in NET format; PARSER-2 adds all the removed hard-blocks into this netlist. It also removes all the previously added inputs and outputs. This final netlist in NET format, containing CLBs and hard-blocks, is passed to PLACER and ROUTER. In future instead of SIS, we intend to use ABC [Berkeley Logic Synthesis and Verification Group, 2005]. Few of the major components of the software flow are described as below.

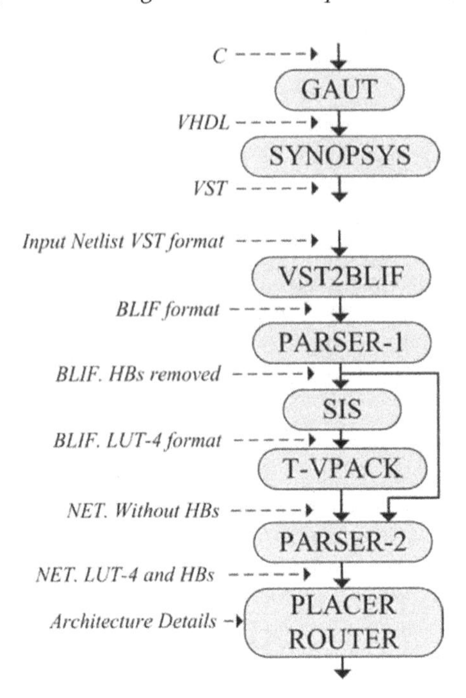

Figure 3.4: Software Flow

3.4.1 Parsers

The output generated by VST2BLIF tool is a BLIF file containing input and output port instances, gates belonging to a standard cell library, and hard-block instances (which are represented as sub circuits in BLIF format). This BLIF file is passed to SIS [E.M.Sentovich et al., 1992] for synthesis into LUT format. However, the hard-blocks in the BLIF file are not required to be synthesized. So, the main aim of PARSER-1 is to remove hard-blocks from BLIF file in such a way that all the dependence between the hard-blocks and the remaining netlist is preserved. After synthesis and packing, PARSER-2 will add all the removed hard-blocks in the netlist.

Figure 3.5 shows five different modifications performed by PARSER-1 before removing hard-block instances from the BLIF file. These cases are described as below :

1. Figure 3.5(a) shows a hard-block whose output pins are connected to the input pins of gates. All the output pins of hard-block are detached from the input pins of gates. These detached signals are added as the input pins of main circuit, as shown in Figure 3.5(b)
2. All the output pins of main circuit that are connected by the output pins of hard-block (as shown in Figure 3.5(c)) are connected to zero gates (as shown in Figure 3.5(d)) This is because, when hard-block is removed, these main circuit outputs do not remain stranded.

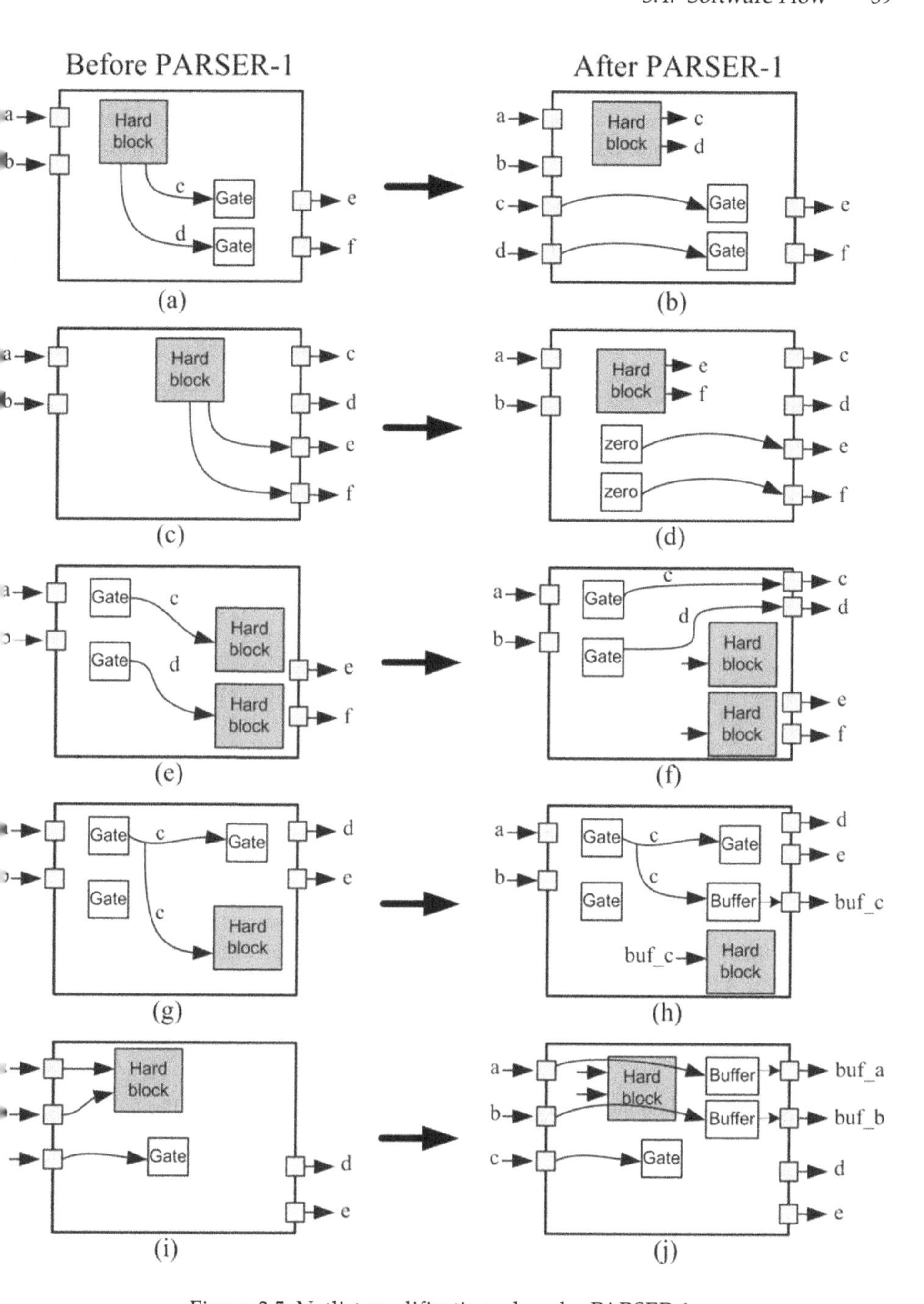

Figure 3.5: Netlist modifications done by PARSER-1

3. All the output pins of gates connected only to the input pins of hard-blocks (shown in Figure 3.5(e)) are added as the output pins of main circuit (as shown in Figure 3.5(f)).
4. For all the output pins of gates connected to the input pins of hard-blocks and also to the input pins of gates (as shown in Figure 3.5(g)), add a buffer to this gate output. The buffered output is added as the output of main circuit. The name of the buffered output should be replaced in all the input pins of hard-blocks. (as shown in Figure 3.5(h))
5. Figure 3.5(i) shows the input pins of main circuit that are connected only to input pins of hard-blocks. After the removal of hard-blocks these inputs will remain stranded and will eventually be removed by SIS. To avoid their removal, these input pins are retained by adding buffers to them, and adding the buffered outputs to the main circuit outputs.

After performing the above changes, PARSER-1 removes all the hard-blocks from the BLIF file. The BLIF file without hard-blocks is passed to SIS, which converts them to LUTs and Flip-Flops. T-VPACK packs LUTs and Flip-Flops together into CLBs. Next, PARSER-2 performs the following changes

1. Adds hard-blocks in the netlist file which is generated by T-VPACK.
2. Removes the "Main circuit input" and "Main circuit outputs" added by PARSER-1.
3. Removes all the zero gates (represented as a clb after SIS) added by PARSER-1.

The final output file contains I/O instances, CLB instances and hard-block instances. This file is passed to placer and router.

3.4.2 Placer

The placement algorithm determines the position of heterogeneous block instances of a netlist on their respective block types on FPGA architecture. The main goal is to place connected instances near to each other so that minimum routing resources are required to route their connections. The placer uses simulated annealing algorithm [S.Kirkpatrick et al., 1983] [C.Sechen and A.Sangiovanni-Vincentelli, 1985] to achieve a placement having minimum sum of half-perimeters of the bounding boxes of all the nets. This placer also optimizes floor-planning of different blocks on the FPGA architecture. A general introduction of simulated annealing based placement algorithm is already described in section 2.1.3. Some of the major changes required for the placement of heterogeneous blocks are described as below :

Bounding box formation :

The PLACER uses bounding box based cost function to optimize the placement of netlists on FPGA architectures. The bounding box of a signal or net is the minimum box which encompasses the driver and the receiver pins of the net. This work considers the CLASS, position and direction of pins in the formation of bounding box (BBX).

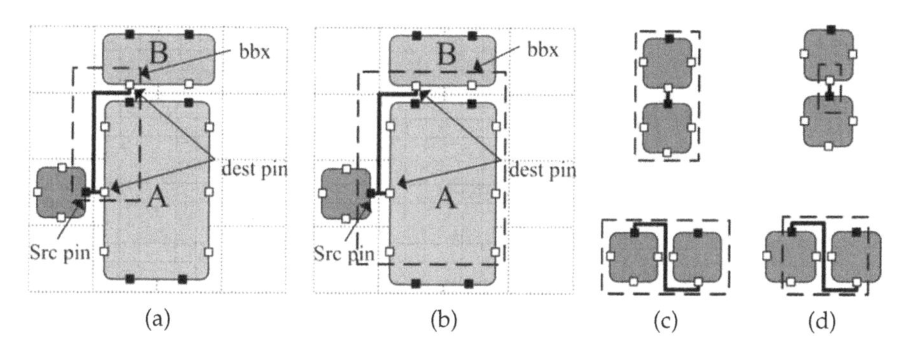

Figure 3.6: Bounding Box

Pins having same CLASS are considered logically equivalent; router can route a net to any of the input pins of a BLOCK having the same CLASS number. For that purpose, the BBX of a net, connected to a receiver pin of a BLOCK, includes all the input pins of a BLOCK having same CLASS number. Thus the bounding box of a net used in this work is the minimum box which encompasses the driver pin and the receiver pins of the net, and all the input pins of a BLOCK having the same CLASS as that of the receiver pin of the BLOCK that is connected to the net. The PLACER tries to achieve a placement having the minimum sum of the half-perimeters of the bounding box of all the nets. Figure 3.6(a) shows a case in which all the input pins (in white color) of BLOCK 'A' are defined to have different CLASS, whereas in Figure 3.6(b) all its input pins are defined to have same CLASS. The bounding box size in Figure 3.6(b) increases as compared to bounding box size in Figure 3.6(a). This increase in the BBX size is due to the inclusion of input BLOCK pins having the same CLASS.

This increase in bounding box size gives an accurate and consistent bounding box cost value, irrespective of the input pin to which the net will be routed. If all the input pins having same CLASS number are not included in the bounding box of a net, such a bounding box will be minimum only if the driver instance driving the net is placed near to a single input pin of the receiver block which is included in the bounding box. The PLACER will also naturally try to place such a driver block near to a single input pin, without knowing that placing the same driver instance near other input pins might also decrease the routing congestion. Thus the bounding box that includes all the input pins of a block having the same CLASS gives multiple placement options to the PLACER, thus improving the quality of placement solution.

This work also considers the position and direction of pins in the formation of bounding box of a net. Figure 3.6(c) shows two cases in which the bounding box is formed without considering the pin positions and directions. The bounding box is only formed by the position of block, and not by the position of pins on the block. Both the top and bottom placement in Figure 3.6(c) have got the same bounding box sizes. However if the amount of routing wires are considered, the top placement requires less routing resources than the bottom placement. It means that a bounding box which does not consider the pin position and directions might not favor a case which requires less routing routing resources. Figure 3.6(d) shows the same

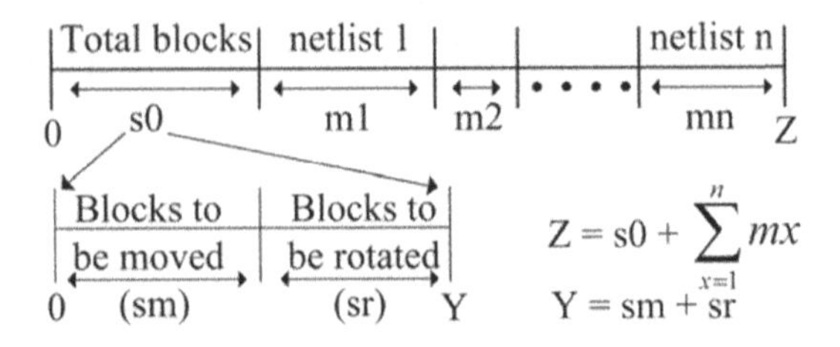

Figure 3.7: Source Selection

two cases; the bounding box is formed here with pin positions and directions taken into consideration. In this case, it can be seen from the sizes of BBX that the top placement is preferred over the placement on bottom.

Placer Operations :

The PLACER either moves an instance from one BLOCK to another, moves a BLOCK from one slot position to another, rotates a BLOCK at its own axis, or moves an entire column of BLOCKS. After each operation, the bounding box cost (also called as placement cost) is re-computed for all the disturbed signals. Depending on the cost value and the annealing temperature, the simulated annealing algorithm accepts or rejects the current operation. Multiple netlists are placed together to get a single architecture floor-planning for all the netlists. For multiple netlist placements, each BLOCK allows multiple instances to be mapped onto it; but multiple instances of the same netlist cannot be mapped on the same BLOCK.

The placer performs its operation on source and destination. The selection of source and destination is described below.

- **Source selection**

  A "source" can be an instance belonging to any of the input netlist or a BLOCK of the FPGA architecture. The probability of selecting a particular "source" is proportional to the ratio of number of total elements of that particular source to the total sum of the elements of all the sources. Thus all the BLOCKS in the architecture, and the instances of all the given set of netlists are given a unique ID as shown in figure 3.7. Z is the sum of the BLOCKS in the architecture that can be moved or rotated and, the instances found in all the input netlists. A random number selected from Z decides if operation is to be performed on a BLOCK or on an instance of any netlist. If the random number represents an instance, the instance of the netlist referred by this random number is chosen as the "source". If it is a BLOCK, then another random operation decides if BLOCK should be moved or rotated. Y is the sum of all the BLOCKS that can be either moved or rotated. The I/O BLOCKS can neither be moved nor rotated. The remaining BLOCKS in the architecture can be moved, and are represented as "sm". The rotation is allowed only for those BLOCKS which have allowed so in the architecture definition of their respective blocks. "sr" is the sum of BLOCKS in the architecture that can be moved. It is to be noted that Y is not equal to "s0" because there can be BLOCKS which can be moved as well as rotated. Thus a random operation on Y selects a BLOCK to be moved or rotated.

- **Destination selection**

  After the selection of "source", a destination is selected where this "source" can be moved. If the source is an instance, then any random matching BLOCK is selected as its destination. If the source is a BLOCK, then a slot position is selected as its destination. If a source BLOCK is to be rotated, the same source slot position becomes the destination. If the source BLOCK is to be moved, then any random slot is selected as its destination. This destination slot selection for BLOCK movement is done using the following steps

  1. Get the size of the source BLOCK. The rectangular window occupied by the source BLOCK is called as the source window. The source window, depicted in dashed line, is shown in figure 3.8(a).
  2. Choose any random slot as destination. The rectangular window starting from the destination slot, having the same size as the source window is called as the destination window. The source window will always contain a single BLOCK, whereas the destination window can contain one or more BLOCKS. A valid destination window depicted in solid line is shown in figure 3.8(b).
  3. The destination slot is rejected if the source and the destination windows do not overlap, and the destination window contains at least one such BLOCK which exceeds the limits of the destination window. Figure 3.8(c) depicts this case. This is done because it would not be possible to move all the BLOCKS in the destination window to the source window.
  4. The destination slot is rejected if the destination window exceeds the boundaries of the slot-grid (as shown in Figure 3.8(d)).
  5. If the source and the destination windows overlap horizontally or vertically, then accept this destination slot position. Horizontal or vertical translation operation will be applied in these cases. Figures 3.8(e) shows a valid destination window overlapping vertically with the source window. Figures 3.8(f) shows the positions of source and destination BLOCKS after vertical translation.
  6. The destination slot is rejected if the source and destination windows overlap diagonally (as shown in Figure 3.8(g)).
  7. If a slot is rejected, the procedure is repeated until a valid destination slot is found.

After the selection of source and destination, any one of the following operations is performed by the placer :

- **Instance Move**

  In this case, a move operation is applied on the source instance and the destination BLOCK. If the destination BLOCK is empty, the source instance is simply moved to the destination BLOCK. If the destination BLOCK is occupied by an instance, then an instance swap operation is performed.

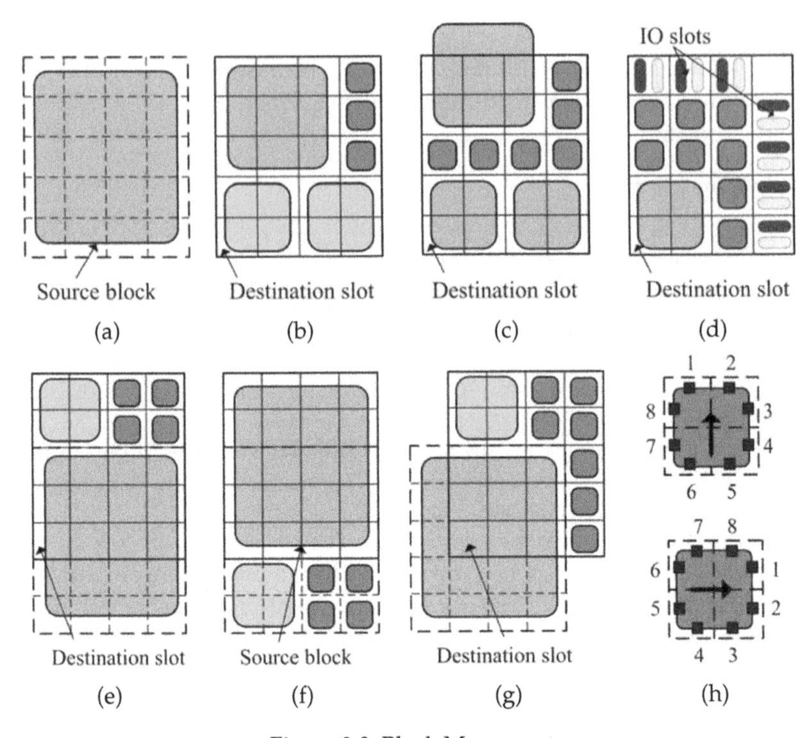

Figure 3.8: Block Movement

- **BLOCK Jump**

  If the source window does not overlap with the destination window, then a JUMP operation is performed. All the BLOCKS in the destination window are moved to the source window, and the source BLOCK is moved in the destination window. The instances mapped on a BLOCK also move along with the BLOCK.

- **BLOCK Translate**

  If the source and the destination windows overlap, then a translation operation is performed. Only horizontal and vertical translations are currently performed, diagonal translation is not performed in this work. Figure 3.8(e) and 3.8(f) show a case of vertical translation. The five BLOCKS found in the upper 2 rows of the destination window (as shown in Figure 3.8(e)) are moved to the lower 2 rows of the source window (as shown in Figure 3.8(f)). The source BLOCK is then moved to the destination window. Figure 3.8(g) shows a source window diagonally overlapping a destination window. Such kind of overlap requires a diagonal translate operation, which is left for future work.

- **BLOCK Rotate**

  The rotation of BLOCKS is important when the CLASS assigned to the input pins of a

BLOCK are different; bounding box varies depending upon the pin positions and their directions. A BLOCK can have an orientation of 0°, 90°, 180° or 270°. The orientation of a BLOCK is used by the bounding box evaluation function to correctly calculate the exact position and direction of each of its PINs. When an instance of a netlist is moved from one BLOCK to another BLOCK having different orientations, the orientation of both the old BLOCK and the new BLOCK are used to compute the difference in the bounding box. Figure 3.8(h) depicts a 90° clock-wise rotation. Multiples of 90° rotation are allowed for all the BLOCKS having a square shape, whereas at the moment only multiples of 180° rotation are allowed for rectangular (non-square) BLOCKS. A 90° rotation for non-square BLOCKS involves both rotation and move operations, which is left for future work.

- **Column Move**

  The column move operation moves a complete column of BLOCKS from one slot position to another. If the source BLOCK is restricted to remain in a column, a column move operation is performed.

3.4.3 Router

After the placement of netlist on the FPGA architecture, the exploration environment constructs routing graph for the architecture. Few of the architecture description parameters required for the construction of routing graph are taken from the architecture description parameters. These parameters mainly include the type of routing network (unidirectional or bidirectional), channel width, I/O rate, block types and their pin positions on the block. Other parameters depend on the floor-planning details. These parameters include the position of blocks on the slot-grid and their orientation (0°, 90°, 180° or 270°). After the construction of routing graph, the PathFinder routing algorithm [L.McMurchie and C.Ebeling, 1995] is used to route netlists on the routing architecture. In case a binary search operation is used, routing graph is constructed for varying channel widths; routing is tried for each channel width until a minimum channel width is found. The routing algorithm described in section 2.1.3 does not require any major changes to route netlists on heterogeneous FPGA architectures.

3.5 Floor-planning Techniques

Various manual and automatic architecture description parameters can be used to generate different heterogeneous architecture floor-plannings. Experiments are performed in this work by using six major floor-plannings as shown in figure 3.9. These floor-plannings are shown for a hypothetical architecture which consists of I/O pads, soft-blocks, and two different kind of hard-blocks. Three floor-plannings in Figure 3.9(a), (b) and (c) are acquired manually, whereas the floor-plannings in Figure 3.9(d), (e) and (f) are automatically manipulated for the given netlists. These floor-plannings are explained as follows.

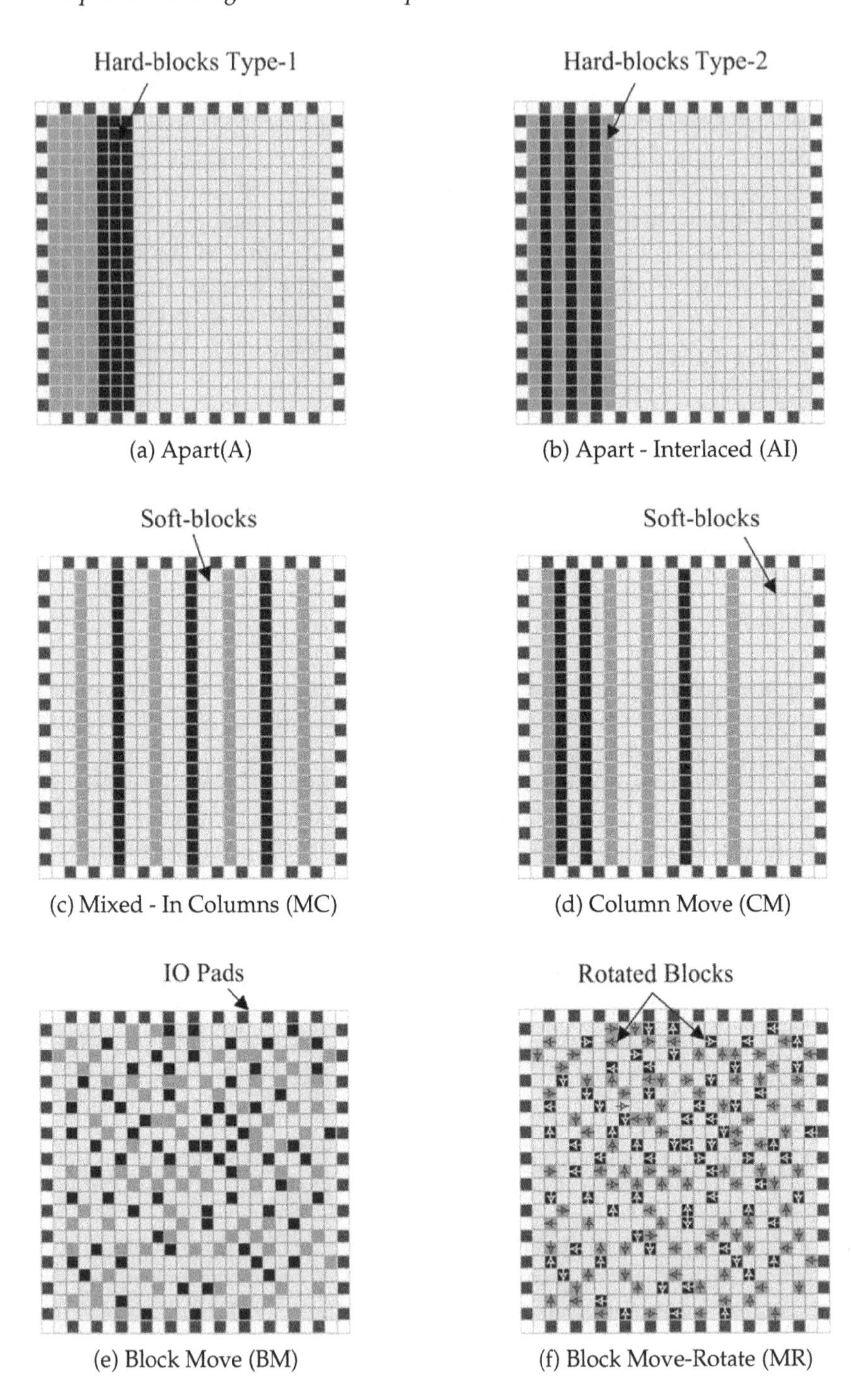

(a) Apart(A)

(b) Apart - Interlaced (AI)

(c) Mixed - In Columns (MC)

(d) Column Move (CM)

(e) Block Move (BM)

(f) Block Move-Rotate (MR)

Figure 3.9: Different Floor-Plannings

Cell name	Area λ^2
Flip-flop	90 x 50
Mux 2:1	45 x 50
Mux 3:1	65 x 50
And 2:1	25 x 50
Or 2:1	25 x 50
Sram	30 x 50
Buffer	20 x 50
Zero	15 x 50
Inv	15 x 50

Table 3.3: Standard cells area

Figure 3.9(a) shows a floor-planning in which heterogeneous hard-blocks are placed apart from the soft-blocks. Such kind of floor-planning can be beneficial for FPGAs optimized for datapath circuits [Cherepacha and Lewis, 1996]. Figure 3.9(b) shows a floor-planning in which hard-blocks are placed apart from the soft-blocks, but columns of different hard-blocks interlace each other. In Figure 3.9(c), hard-blocks blocks are arranged in columns, and are placed at equal distance with one another. The columns of hard-blocks are separated by columns of soft-blocks. This type of floor-planning is employed by current state-of-the-art commercial FPGA architectures [Virtex, 5] [Stratix, IV]. Figure 3.9(d) shows a floor-planning in which the columns of hard-blocks optimize their positions for the given set of netlists. Figure 3.9(e) shows a floor-planning in which different blocks optimize their positions for the given set of netlists. In Figure 3.9(f), blocks optimize their position and orientation for the given set of netlists.

Floor-planning can have a major implication on the area of an FPGA. If a tile-based layout is required for an FPGA, similar blocks can be placed in same column. In this way, width of the entire column can be adjusted according to the layout requirements of the blocks placed in a column. On the other hand, if blocks of different types are placed in a column, the width of the column cannot be fully optimized. This is because the column width can only be reduced to maximum width of any tile in that particular column. Thus, some unused space in smaller tiles will go waste. Such a problem does not arise if a tile-based layout is not required. In such a case, an FPGA hardware netlist can be laid out using any ASIC design flow.

However, column-based floor-planning restricts the number of hard-blocks in an FPGA to be in multiples of hard-blocks that can fit in a column. Thus, some hard-blocks might remain unused. Similarly, communication between different hard-blocks that are placed in different columns might require additional routing resources; thus FPGA area increases. This work explores differences in placement costs, minimum channel widths, and eventually total area of heterogeneous FPGA for varying floor-plannings.

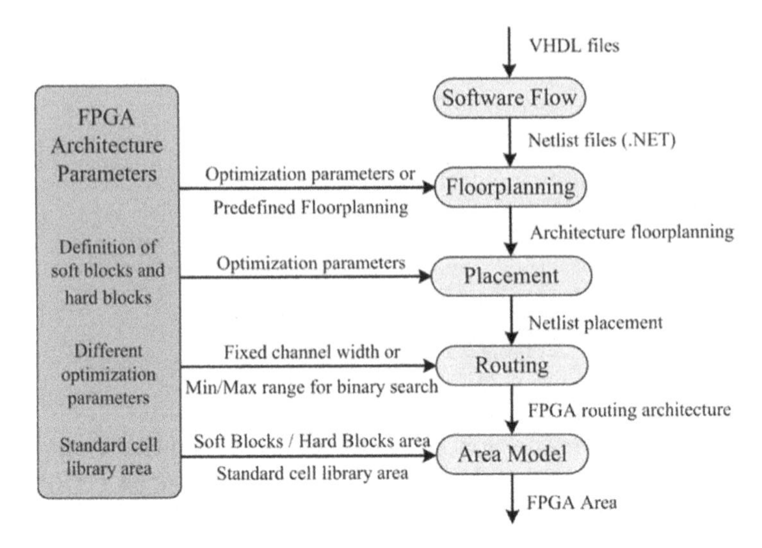

Figure 3.10: Experimental Flow

3.6 Area Model

A generic area model is used to calculate the area of the FPGA. The area model is based on the reference FPGA architecture shown in Figure 3.2. Area of SRAMs, multiplexers, buffers and Flip-Flops is taken from a symbolic standard cell library (SXLIB [Alliance, 2006]) which works on unit Lambda(λ). Area of few of the cells of standard cell library are shown in Table 3.3. The area of FPGA is reported as the sum of the areas taken by the switch box, connection box, buffers, soft logic blocks, and hard-blocks.

3.7 Experimentation and Analysis

Different set of test-bench circuits are used to show the effect of floor-planning on the area of heterogeneous FPGA architectures. This work employs two experimental methodologies. In the first methodology, a group of netlists are mapped on a common FPGA architecture. In the second methodology, FPGA architecture is optimized individually for each netlist, average results for all the netlists are presented. These two experimental methodologies, test-bench circuits used for experiments and, their results and analysis are presented as below.

3.7.1 Experimental Methodology using Common Architecture

In this methodology, a common FPGA architecture is defined that can map any circuit in the set of test-bench circuits. These test-bench circuits are initially converted to NET format through a software flow described earlier. The group of netlists are simultaneously placed on the architecture and their floor-planning is refined. Once the final floor-planning is achieved, each netlist is individually placed on it (without changing the floor-planning). Experiments are done using six different floor-plannings that are described in the previous section. After placement, each netlist is routed on the architecture. A single-driver unidirectional routing network is used for all experiments, and maximum routing iteration is set to 50. After routing, area of FPGA is computed through an area model. The proposed experimentation flow is illustrated in Figure 3.10.

Experiments are performed with normal placement time, and with enhanced placement times to ensure best possible results. Generally, netlists are placed on the architecture using PLACER which uses a bounding-box based simulated annealing algorithm. The simulated annealing algorithm decreases the temperature gradually; a fixed number of placer operations are attempted in each temperature iteration. The total number of operations in a temperature step are directly proportional to the total time taken by the PLACER. Increasing the total number of operations also improves the placement quality. However, as the simulated annealing based algorithm is based on heuristics, increasing the number of operation does not guarantee to find the best possible result. Experiments are performed with normal placement times, and with 50X placement time. The main purpose of performing the experiments with more execution time is to ensure that best achievable placement is attempted for each floor-planning.

When a common FPGA architecture is used to map a set of netlists, the architecture size, routing channel width, and most importantly floor-planning of the maximum architecture are sometimes influenced by the largest netlist in the group of netlists. So, experiments are also performed by optimizing the floor-planning only for the single largest netlist in the group. Other netlists in the group are placed on the FPGA floor-planning that is optimized for largest netlist.

3.7.2 Benchmark Circuits for Common Architecture

Experiments are performed on three different sets of circuits. The first set of test-bench is shown in Table 3.4. It comprise of 4 benches which are a subset of 8-bit FIR, FFT, ADAC, and DCU. These are relatively small benches, and are obtained from netlist generators written in a procedural language, STRATUS. These netlists contain CLBs along with 5 different types of hard-blocks. The second set of 3 benches, shown in Table 3.5, are generated from GAUT [GAUT, 2010]. These are relatively large benches. The CLBs in these benchmarks comprise of a cluster of four BLEs. These netlists contains 2 different type of hard-blocks (i.e. multipliers and adders). Third set of test-benches, shown in Table 3.6, comprise of 10 open-core netlists [OpenCores, 2010]. These netlists also contain 2 different type of hard-blocks. Different soft-blocks and hard-blocks used by the three sets of netlists are initially defined in the architecture description file. All the hard-blocks that occupy more than one slot allow routing channel to pass through them. A maximum FPGA architecture is defined for each set of circuits. The composition of a maximum architecture for each set of netlists is shown in the last row of Table 3.4, 3.5 and 3.6.

Index	Netlist Name	No. of Inputs	No. of Outputs	No. of CLBs	No. of Mul_8_8_16	No. of Slansky_16	No. of Sff_8	No. of Sub_8	No. of Smux_16
1.	Fir	9	16	32	4	3	4	-	-
2.	Fft	48	64	94	4	3	-	6	-
3.	Adac	18	16	47	-	-	2	-	1
4.	Dcu	18	16	34	1	1	4	2	2
	Maximum	48	64	94	4	3	4	6	2

Table 3.4: Netlist-1 block utilization table - Stratus Netlists

Index	Netlist Name	No. of Inputs	No. of Outputs	No. of CLBs (Cluster of 4)	No. of Mul_16_16_16	No. of Add_16_16_16
1.	Fir16	259	16	572	16	8
2.	Prodmat	291	80	1112	27	11
3.	Ellipticass	67	48	818	-	15
	Maximum	291	80	1112	27	15

Table 3.5: Netlist-2 block utilization table - GAUT Netlists

Index	Netlist Name	No. of Inputs	No. of Outputs	No. of CLBs	No. of Mult_16x16	No. of Add_20x20
1.	cf_fir_24_16_16	418	37	2149	25	48
2.	rect2polar	34	40	1328	0	52
3.	polar2rect	18	32	803	0	43
4.	cfft16x8	20	40	1511	0	26
5.	fm_receiver	10	12	910	1	20
6.	fm_receiver2	9	12	1308	1	19
7.	cf_fir_7_16_16	146	35	638	8	14
8.	lms	18	16	940	10	11
9.	rs_encoder	138	128	537	16	16
10.	cf_fir_3_8_8	42	18	159	4	3
	Maximum	418	128	2149	25	52

Table 3.6: Netlist-3 block utilization table - Opencores Netlists

3.7.3 Results and Analysis for Common Architecture

Experimental results for three groups of netlists are presented individually. Results for Netlist-1 are shown in Figure 3.11. Three graphs are shown, Figure 3.11(a) with normal placement time, Figure 3.11(b) with 50X placement time, and Figure 3.11(c) with 50X placement time with floor-planning optimized only for the maximum netlist in the group. In each of these graphs, X-axis shows the results for six different floor-planning techniques. The left Y-axis represent the placement costs for different netlists in the group, and the total sum of placement cost of all the netlists. The right Y-axis represent the maximum channel width required by the architecture to route any of the netlists in the group. Figure 3.11(a) shows 30% total placement cost difference between the worst and the best floor-planning (i.e. between "Apart" and "Block Move and Rotate"). The difference in placement cost gives 14.2% advantage in channel width (i.e. decreases the channel width from 14 to 12). Since, major area of an FPGA is occupied by the routing network, 14.2% benefit in routing channel width gives an overall 11.6% area advantage (area results are not displayed in the graphs). The best floor-planning compared to "Interlaced/In Column" floor-planning shows 18% placement cost, 14.2% channel width and 11.6% overall area advantage.

The results in Figure 3.11(b) are taken with 50X more placement time than the normal time. The main aim is to show difference in placement cost when placement time is increased by 50 times. It can be seen in the Figure that only slight placement gains are achieved by increasing the placement time. The slight placement gains have been able to reduce the maximum channel width of the floor-planning achieved through Column Move operation. For other floor-plannings, the routing channel width remains unchanged.

Figure 3.11(c) shows experimental results for floor-plannings that are only optimized for FFT. It is to be noted that floor-planning is optimized only for "Column Move", "Block Move" and "Block Move and Rotate". The remaining three floor-plannings are fixed, and are independent of the netlists. It can be seen in Figure 3.11(c) that the placement cost of FFT is significantly reduced for "Block Move" and "Block Move and Rotate", whereas the total placement cost has increased. This is because the floor-planning that is optimized for FFT only is not the best floor-planning for other netlists, and thus placement cost increases for other netlists. But, eventually the maximum channel width reduces to 10 for "Block Move and Rotate" floor-planning. Thus 29% improvement in channel width and, 28% improvement in total area is recorded between an "Interlaced/In Column" floor-planning and "Block Move and Rotate" floor-planning.

Experimental results for Netlist-2 are shown in Figure 3.12. Figure 3.12(a) shows 31% placement cost difference, 62% channel width difference and eventually 51% total area difference between the worst and best floor-planning (i.e. between "Apart" and "Block Move and Rotate"). There is a very slight placement cost difference between the best floor-planning and "Interlaced/In Column" floor-planning, there is no channel width gain, and thus area remains the same. However, as the placement time is increased as shown in Figure 3.12(b), 7.5% difference in channel width is recorded between "Block Move and Rotate" and "Interlaced/In Column" floor-planning; thus 5% area gain is achieved. Figure 3.12(c) shows no major channel width gain when floor-planning is optimized for the maximum netlist in the group.

Experimental results for Netlist-3 are presented in Figure 3.13. Figure 3.13(a) shows 8.5% placement cost different, 25% channel width difference, and eventually 21.5% area gain between the "Interlaced/In Column" and "Block Move and Rotate" floor-planning. When floor-planning is optimized only for the maximum netlist in the group, placement cost difference increases to 15%, channel width difference increases to 37.5% and eventually total area difference increases to 35.5% between the "Interlaced/In Column" and "Block Move and Rotate" floor-planning.

The above results help us draw the following conclusions

- The "Apart" floor-planning is not a suitable floor-planning; atleast not for the netlists used in this work. This floor-planning might be advantageous (i) if control-path portion of a circuit implemented on CLBs is relatively small as compared to data-path portion of circuit implemented on hard-blocks, and (ii) if routing network of control- and data-path sections of the FPGA architecture are optimized independently.
- Column based floor-planning of hard-blocks is advantageous for an optimized tile-based layout generation; the widths of hard-blocks placed in columns can be appropriately adjusted to optimize the layout area. However, column-based floor-planning is unable to decrease the placement cost as few other floor-plannings do. This difference in placement costs can sometimes result in as high as 35.5% difference in total area of FPGA.
- The floor-planning achieved through Block Move and Rotate operation gives the least possible placement cost, and eventually least FPGA area as compared to other floor-plannings. However, such a floor-planning can be achieved only if the set of netlists are known in advance. Such can be a case if an application specific FPGA is desired for a product.
- Area advantages can be attained if floor-planning is optimized only for netlist(s) that occupy maximum routing channel width.

Floor-planning achieved through block move and rotate operation might not give an area efficient tile-based layout. This is because, the size of hard-block tile must be a multiple of the smallest tile in an FPGA, which is generally a CLB. Thus, some area will eventually be wasted in each hard-block. However, the extra free area in each hard-block can be used to employ some other area saving technique such as shadow clustering [Jamieson and J.Rose 2006]. The shadow clustering technique increases the area efficiency of FPGAs by adding a shadow CLB cluster in a hard-block. If the hard-block is not used by a netlist, the shadow cluster can be used for implementing some general purpose logic. The main aim of a shadow clustering is to utilize the precious routing resources of the hard-block if it remains unused in an FPGA.

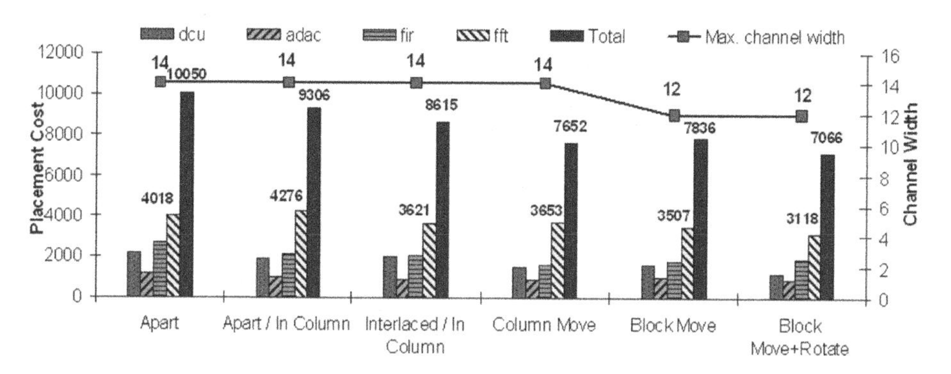

(a) Normal placement - Floor-planning optimized for all netlists

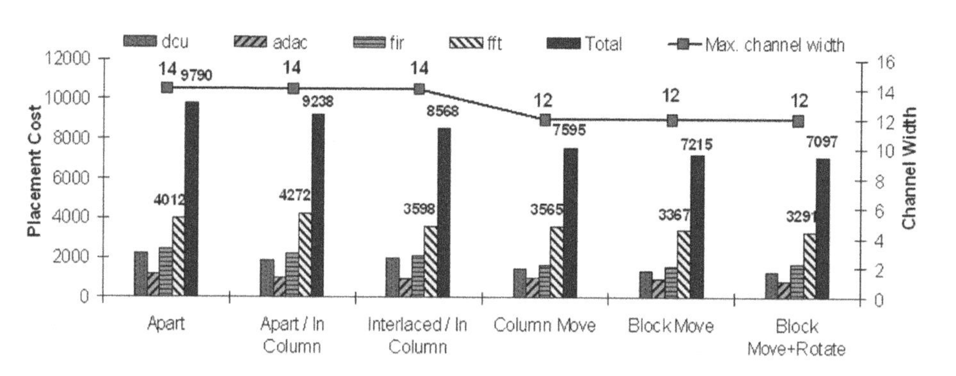

(b) 50x placement time - Floor-planning optimized for all netlists

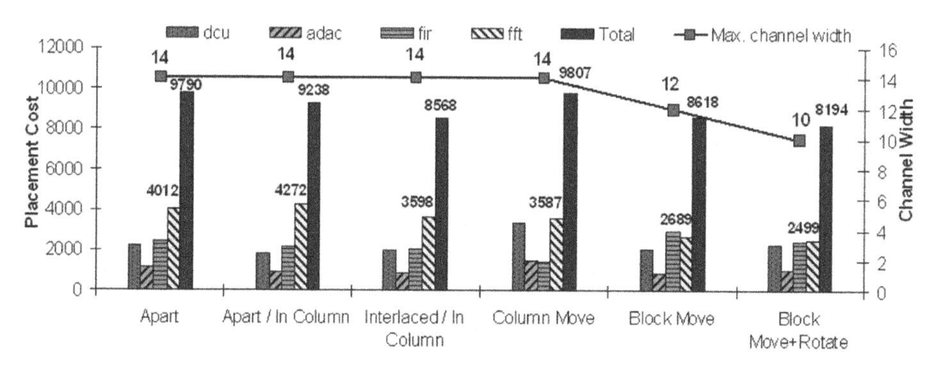

(c) 50x placement time - Floor-planning optimized for largest netlist

Figure 3.11: Experimental results for Netlist-1 (Stratus) using common architecture

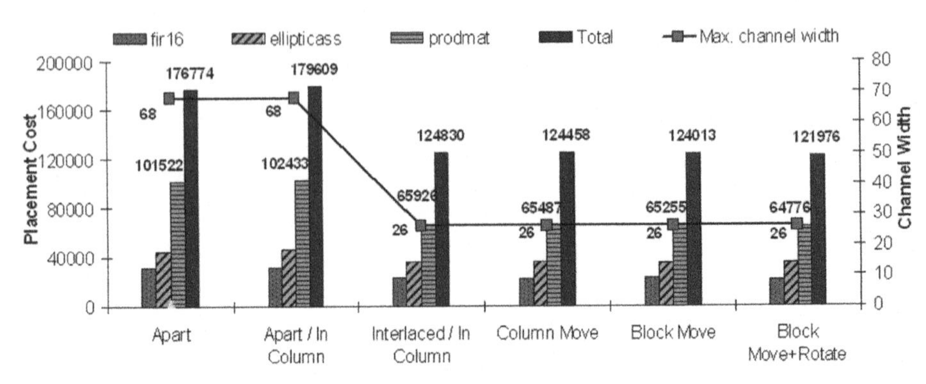

(a) Normal placement - Floor-planning optimized for all netlists

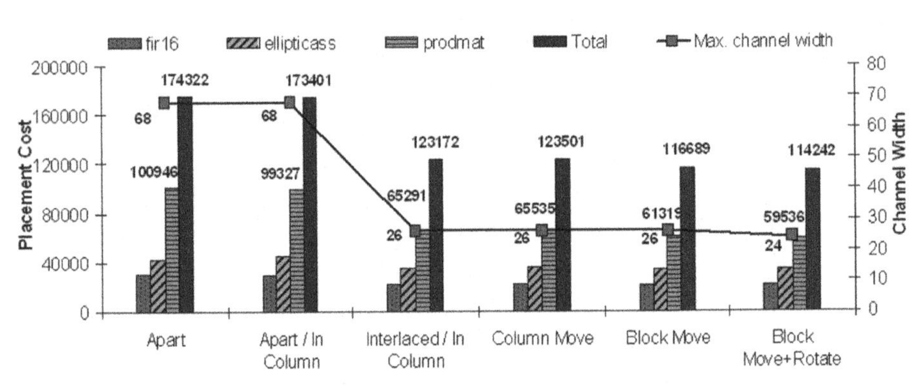

(b) 50x placement time - Floor-planning optimized for all netlists

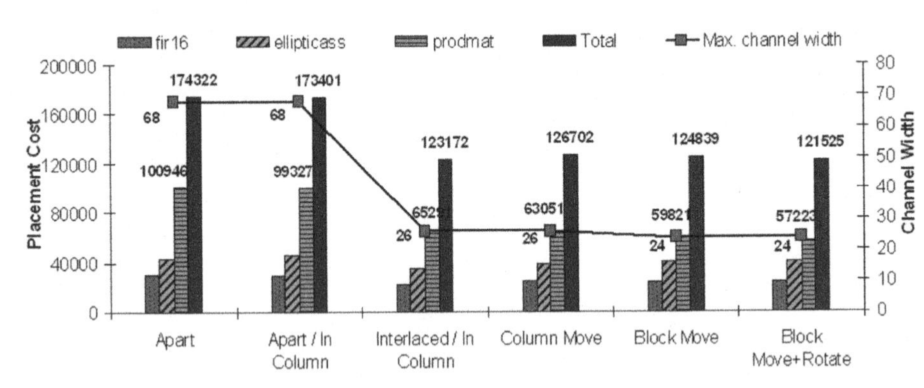

(c) 50x placement time - Floor-planning optimized for largest netlist

Figure 3.12: Experimental results for Netlist-2 (GAUT) using common architecture

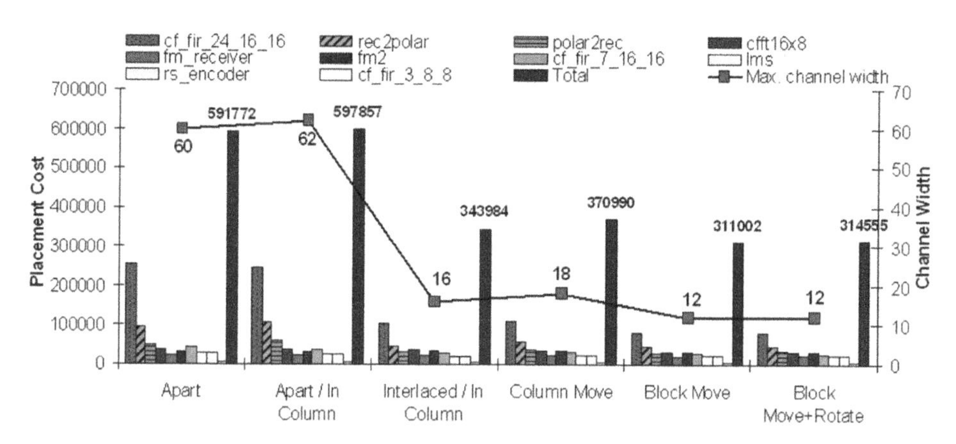

(a) Normal placement - Floor-planning optimized for all netlists

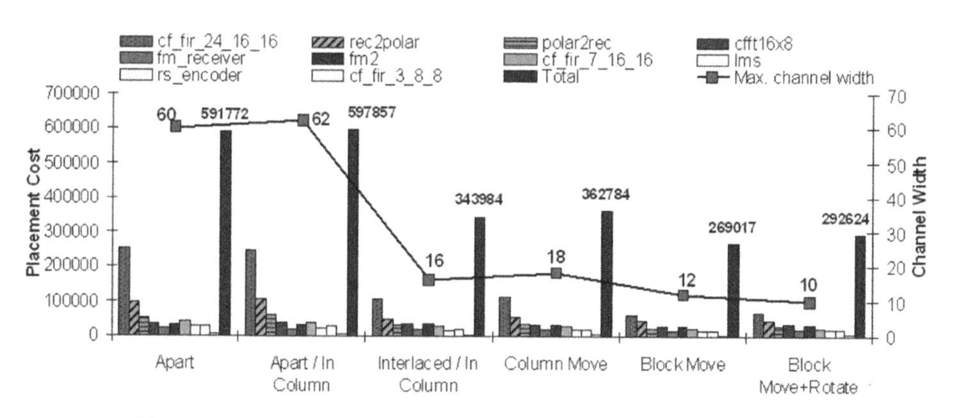

(b) 50x placement time - Floor-planning optimized for largest netlist

Figure 3.13: Experimental results for Netlist-3 (Opencores) using common architecture

3.7.4 Experimental Methodology using Individual Architectures

In this methodology, FPGA architecture and its floor-planning is optimized for each netlist individually. Each netlist is converted to NET format through a software flow described earlier. The floor-planning and the placement is optimized for the netlist. Experiments are done using six different floor-plannings that are described in the previous section. After placement, each netlist is routed on the architecture. A single-driver unidirectional routing network is used for all experiments, and maximum routing iteration is set to 50. After routing, area of FPGA is computed through an area model.

In the previous experimental methodology, a maximum architecture is generated for three groups of netlists. It was noticed that the routing channel width of an architecture is mainly decided by the maximum channel width required by any of the netlist in the group. Similarly, the floor-planning of different blocks on the architecture was also mainly influenced by the maximum netlist in a group of netlists. Thus, the area gain results for three groups of netlists presented in the previous section are roughly the results for three maximum netlists in each of the three groups of netlists. So, in order to get more realistic results, experiments are performed on individual netlists. Instead of using a common architecture for a group of netlists, FPGA architecture and its floor-planning is optimized for each netlist individually. Later, the average results of all the netlists gives a more realistic difference between different floor-planning approaches.

3.7.5 Benchmark Circuits for Individual Architectures

Experiments are performed for 18 different netlists shown in Table 3.7. The netlists are arranged according to different types of hard-blocks used in them. Netlists 1-5 use only one type of hard-block, netlists 6-14 use two types of hard-blocks, and netlists 15-18 use more than two types of hard-blocks. The netlists in Table 3.7 comprise of four distinct groups. The first set of 5 benches are received from the university of Toronto. These netlists use CLBs and Multipliers. The next three groups of netlists are the same as shown in Table 3.4, 3.5 and 3.6. The last 4 netlists in Table 3.7 are the same as in Table 3.4, however the netlist instances are replicated ten times to achieve prominent results.

3.7.6 Results and Analysis for Individual Architectures

Figure 3.14 shows three different graphs, depicting placement costs, channel widths, and area results for different floor-plannings. The six different floor-plannings used in this experiment are as follows (i) "Column" floor-planning: Different hard-blocks are placed in columns at equal distance with one another. Only the number of hard-blocks required by a particular netlist are placed in columns. The remaining space left in columns of hard-blocks is filled by CLBs. (ii) "Column Move" floor-planning: Columns of different heterogeneous blocks are moved on the architecture. (iii) "Column Full" floor-planning: Different hard-blocks are placed in columns at equal distance with one another. Extra-space left in the columns of

Index	Netlist Name	No. of Blks	No. of CLBs	No. of Blk-1	No. of Blk-2	No. of Blk-3	No. of Blk-4	No. of Blk-5
1.	toronto_cf_fir_24_16_16.4	1	3273	25	0	0	0	0
2.	toronto_fir_scu_rtl_for_cmm_exp.4	1	1366	17	0	0	0	0
3.	toronto_paj_raygentop_no_mem.4	1	3742	18	0	0	0	0
4.	toronto_rs_decoder_1.4	1	1553	13	0	0	0	0
5.	toronto_rs_decoder_2.4	1	2960	9	0	0	0	0
6.	open_cores_rec2polar	1	1328	0	52	0	0	0
7.	open_cores_polar2rector	1	803	0	43	0	0	0
8.	open_cores_cff_16x8	1	1511	0	26	0	0	0
9.	open_cores_cf_fir_24_16_16	2	2149	25	48	0	0	0
10.	open_cores_cf_fir_7_16_16	2	638	8	14	0	0	0
11.	open_cores_lms	2	940	10	11	0	0	0
12.	open_cores_rs_encoder	2	537	16	16	0	0	0
13.	gaut_fir16	2	572	16	8	0	0	0
14.	gaut_prodmat	2	1112	27	11	0	0	0
15.	stratus_adac	2	470	0	0	20	0	10
16.	stratus_fft	3	940	40	30	0	60	0
17.	stratus_fir	3	320	40	30	40	0	0
18.	stratus_dcu	5	340	10	10	40	20	20

* For Netlists 1-5, CLB contains 1 BLE (LUT-4), Blk-1 is 18-bit mulitplier
* For Netlists 6-12, CLB contains 1 BLE (LUT-4), Blk-1 is 16-bit multiplier, Blk-2 is 20-bit adder
* For Netlists 13-14, CLB contains 4 BLE (LUT-4), Blk-1 is 16-bit multiplier, Blk-2 is 16-bit adder
* For Netlists 15-18, CLB cotains 1 BLE(LUT-4), Blk-1 is 8-bit multiplier, Blk-2 is 16-bit Slansky, Blk-3 is 8-bit register, Blk-4 is 8-bit subtractor, BLK-5 is 16-bit multiplexor

Table 3.7: Netlist-4 block utilization table

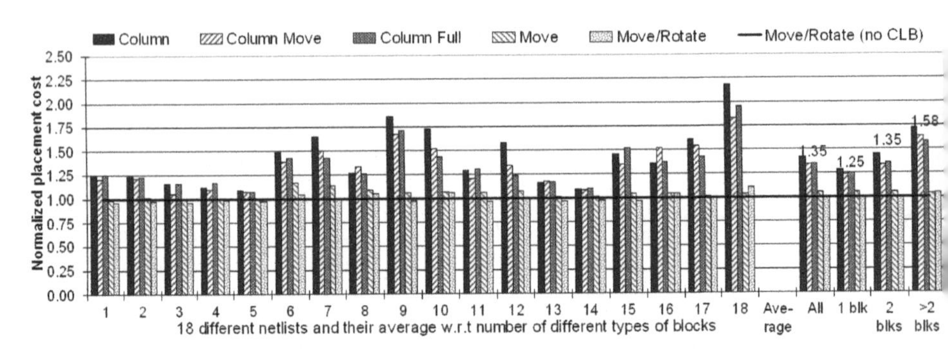

(a) Normalized placement cost to Move/Rotate(no CLB) floor-planning

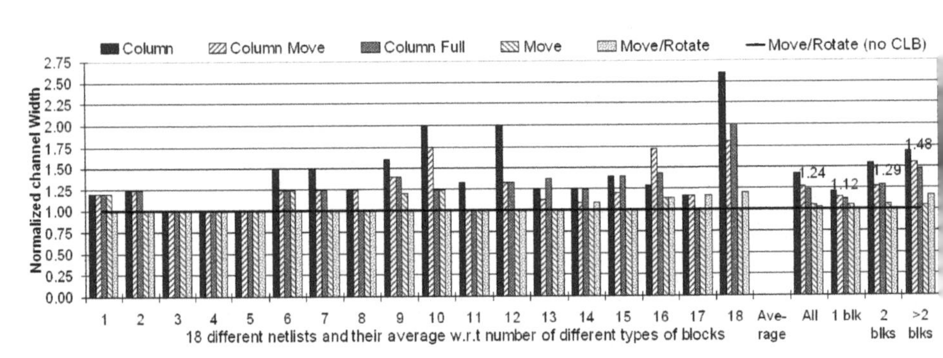

(b) Normalized channel width to Move/Rotate(no CLB) floor-planning

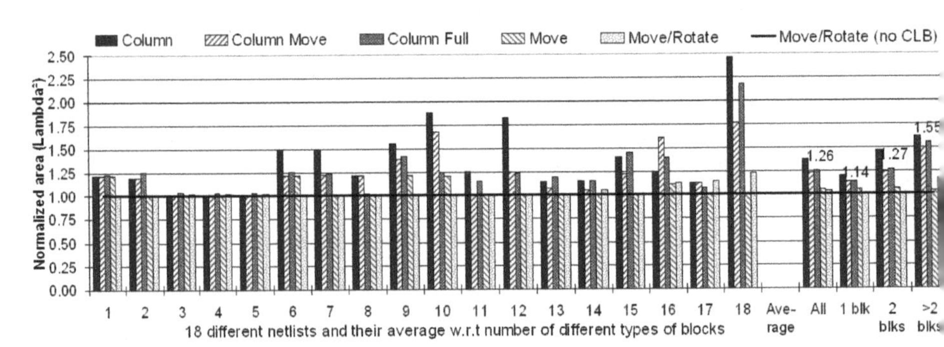

(c) Normalized area (in lambda^2) to Move/Rotate(no CLB) floor-planning

Figure 3.14: Experimental results for Netlist-3

hard-blocks are filled by adding the same type of hard-blocks in the columns. In this floor-planning, special care has been taken while deciding the shape of FPGA. Generally an FPGA shape is selected to be a square. However in some netlists, FPGA shape is slightly modified to avoid adding an entire new column of hard-blocks only for the sake of few hard-blocks. (iv) "Move" floor-planning: Different blocks on the architecture change their positions. (v) "Move/Rotate" floor-planning: Different blocks on the architecture change their position and orientation. (vi) "Move/Rotate (no CLB)" floor-planning: Different blocks on the architecture change their position, while the orientation of only hard-block is changed.

The X-axis in each of the three graphs show 18 netlists mapped on FPGAs for six different floor-plannings. The X-axis also shows the average results of all the netlists, and the average of the netlists according to different types of hard-blocks used by them. The Y-axis in these three graphs show the normalized placement cost, channel width and area results. The results are normalized to Move/Rotate(no CLB) floor-planning.

Figure 3.14(a) shows the normalized placement cost for all the netlists. The "Column Full" floor-planning has on average 7% less placement cost than the "Column" floor-planning. The "Column" floor-planning has exactly the required number of hard-blocks, whereas "Column Full" adds some extra hard-blocks in some columns of the architecture. Although, these extra hard-blocks in "Column Full" floor-planning remain unused. But they also give more placement options for the instances of these hard-blocks. Thus placement cost is reduced. The "Column Full" floor-planning has on average 35% more placement cost than the "Move/Rotate(no CLB)" floor-planning. The "Column Full" floor-planning for netlists having only 1 type of hard-blocks has on average 25% more placement cost, and netlists having more than 2 types of hard-blocks has on average 58% more placement cost than the "Move/Rotate(no CLB)" floor-planning. The "Move/Rotate" floor-planning, (which moves and rotates all the blocks including CLBs) has roughly the same placement cost as "Move /Rotate(no CLB)" floor-planning. The placement cost difference between "Column Full" and "Move /Rotate(no CLB)" floor-planning suggests that limiting the hard-blocks to be placed in columns deteriorates the overall placement cost. This deterioration in placement cost is partly due to the same phenomenon in which the shape of FPGA (square or rectangle) effects the placement cost of a netlist being mapped on FPGA [Betz et al., 1999]. Other reason in the deterioration of placement cost includes increased distance between different blocks placed in different columns. This increase in placement cost requires more routing channels to route signals, and thus area increases.

Decrease in placement cost also decreases routing channel width requirement. Figure 3.14(b) shows the normalized channel width for all the netlists. However, in some cases differences in placement costs are not sufficient enough to decrease channel width requirement e.g. "Move/Rotate (no CLB)" floor-planning is unable to achieve channel width gains over "Column Full" floor-planning for netlists 3, 4, 5, 11 and 17. The "Column Full" floor-planning requires on average 24% more channel width than the "Move/Rotate(no CLB)" floor-planning. It can be noted in Figure 3.14(b) that netlists 14 and 17 require more channel width for "Move/Rotate" floor-planning than for "Move/Rotate(no CLB)" floor-planning. However, the the placement costs for these netlists (shown in Figure 3.14(a)) are less for "Move/Rotate" than "Move/Rotate(no CLB)" floor-planning. This effect is due to the rotation of CLBs in "Move/Rotate" floor-planning. As the bounding box of nets also takes the directions of pins in consideration, the rotation of CLBs sometimes gives much better placement cost; however rotation of CLBs can create more congestion on few routing channels. Increased congestion requires more channel width to route the signals. For that reason, netlist 14 and 17 require more channel width despite having less placement cost for "Move/Rotate" floor-planning compared to "Move/Rotate(no CLB)".

Decrease in channel width decreases the overall area of the FPGA. Figure 3.14(c) shows the normalized FPGA area for all the netlists. The "Column Full" floor-planning takes on average 26% more area than the "Move/Rotate(no CLB)" floor-planning. For the netlists using only 1 type of hard-blocks the average area difference is only 14%, for netlists using 2 types of hard-blocks the average area difference is 26%, and for netlists using more than 2 types of hard-blocks, the average area difference is 55%.

3.8 Conclusion

This chapter presented a new environment for the exploration of heterogeneous FPGA architectures. It presented an architecture description mechanism to define a heterogeneous FPGA architecture. A software flow maps application designs on the architecture. The major feature of this exploration environment is that it can optimize the floor-planning of the architecture for a given set of netlists. Experiments are performed to show the effect of floor-planning on the area of heterogeneous FPGA architectures. It has been shown that on average a column based FPGA takes 26% more area than the best non-column based FPGA. This area difference increases as the number of different types of hard-blocks increase in the FPGA architecture. However, these area gains might decrease due to layout inefficiencies of tile-based non-column heterogeneous FPGAs.

Next chapter proposes a tile-based layout generator for homogeneous FPGA architectures. The homogeneous FPGA layout is generated, validated, and taped-out. Netlist bitstreams are also generated, which are programmed and verified on the generated FPGA architecture. In future, the same generator can be used to generate tile-based heterogeneous FPGA layout.

4

FPGA Layout Generation

This chapter presents an automated method of generating a tile-based FPGA layout. The main purpose of developing a generator is to reduce the overall FPGA design time with limited area penalty. This generator works in two phases. In the first phase, it generates a partial layout using generic parameterized algorithms. The partial layout is generated to obtain a fast bitstream configuration mechanism, an efficient power routing and a balanced clock distribution network. In the second phase, the generator completes the remaining layout using automatic placer and router. This two-phase technique allows better maneuvering of the layout according to initial constraints. The proposed method is validated by generating layout of 1024 (32x32) Look-up Table based mesh FPGA which includes hardware support for the mitigation of Single Event Upsets (SEU) [Elder et al., 1988]. The FPGA layout is generated in a symbolic standard cell library which allows easy migration to any process technology. This layout is successfully migrated and taped-out using 130nm 6-metal layer CMOS process of STMicroelectronics [STMicroelectronics, 2010].

4.1 Introduction and Previous work

Developing a new FPGA is a time consuming and a challenging task. It is reported by [Padalia et al., 2003] that the creation of a new FPGA requires approximately 50 to 200 person years, thus increasing the overall time-to-market of the final product. It is an interesting option to significantly reduce the time-to-market of the product at the expense of limited area penalty. One way to do this is by automating the complete FPGA design process. This chapter discusses automatic layout generation of FPGAs using open-source VLSI tools.

H. Parvez and H. Mehrez, *Application-Specific Mesh-based Heterogeneous FPGA Architectures*,
DOI 10.1007/978-1-4419-7928-5_4, © Springer Science+Business Media, LLC 2011

The generator presented here employs an elegant scheme to integrate manual intervention in the FPGA layout generation procedure. This is done with the help of generic parameterized algorithms which generate a partial layout. The remaining layout is completed by automated layout tools. The partial layout is performed on those portions of the design that are either important in one aspect or other, or are too difficult to be handled properly by automated tools. In this work, the partial layout performs power routing, clock distribution and layout of bitstream configuration mechanism.

A number of previous attempts have been made regarding automated layout generation of FPGAs. One of the major works in this domain is the GILES project (Good Instant Layout of Erasable Semiconductors) [Kuon et al., 2005] [Padalia et al., 2003]. They have demonstrated complete automation of FPGA layout generation with significantly reduced manual labor. The GILES tool [Padalia et al., 2003] automatically generates a transistor-level schematic of an FPGA tile from a high-level architectural specification. The output of GILES is a mask-level layout of a single tile that can be replicated to form an FPGA array. The clock and power segments are later routed using SKILL [SKILL, 2010]. Phillips and Hauck have focused on the automatic layout of domain specific reconfigurable systems [Phillips and Hauck, 2002]. They have reduced the configurability required by an application domain, and thus have generated smaller layouts.

These previous FPGA generators have used commercial VLSI tools; whereas this work presents an FPGA generator based solely upon open-source VLSI tools. The open-source tools can be easily adapted for any specific demands. This work also defines a set of layout parameters to tailor the layout according to the desired requirements. Moreover, FPGA layout is generated in a symbolic standard cell library which allows easy migration to any process technology.

The layout generation methodology presented in this work is an enhancement of the previous work done at LIP6 regarding the fabrication of a 8x8 mesh-based redundant FPGA core [Mrabet, 2009].

4.2 FPGA Generation

This work focuses on tile-based mesh FPGA layout generation. The mesh FPGA architecture under study comprises of an array of configuration logic blocks (CLBs). Each CLB contains a 4 input Look-Up Table (LUT-4) followed by an optional Flip-Flop. Each CLB has 4 inputs (one on each side) and an output that derives adjacent channels on its top and right sides. The CLBs communicate with each other through a disjoint bidirectional routing network. All the inputs and outputs of a CLB connect with all the wires in a channel (i.e. Fc=1). The generated FPGA matrix can have 'Nx' CLBs in X direction, 'Ny' CLBs in Y direction, and a channel width 'Ch'.

The FPGA generator generates both the netlist and the layout of an FPGA. The netlist and layout generation flow of an FPGA is shown in Figure 4.1. An FPGA is divided into different repeatable modules, called as tiles (explained in detail in section 4.2.2). A tile generator

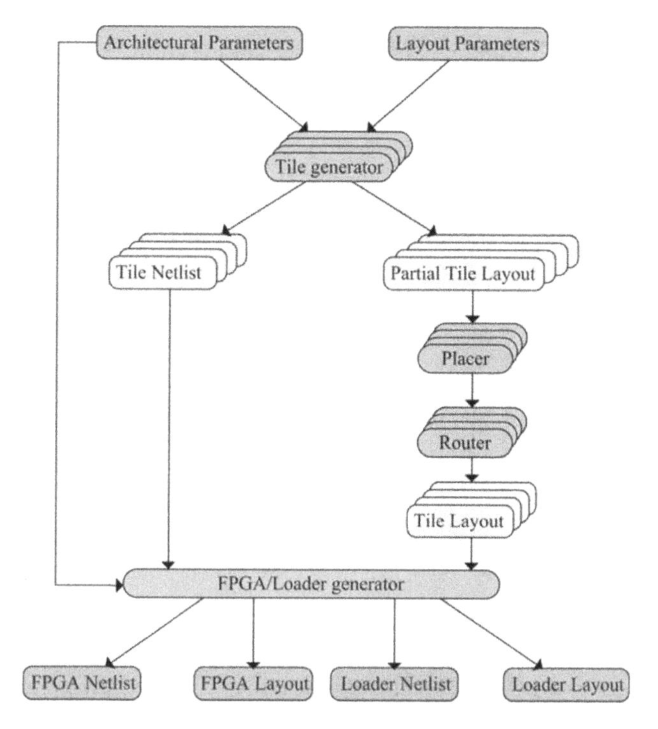

Figure 4.1: Complete FPGA generation CAD flow

initially generates netlist and partial layout of an FPGA tile. The tile generators of an FPGA receive a set of architectural and layout parameters as input. It then generate the netlist and partial layout of tiles in accordance with the given parameters. The partial layout of tiles is generated to obtain a regular configuration memory layout, an efficient power routing and a balanced clock distribution network. The remaining layout of tiles is completed using automatic placer and router. The netlist and layout of all the tiles are abutted together by an FPGA generator to generate the netlist and layout of the complete FPGA. The FPGA generator also generates the netlist and layout of a hardware module named as loader. The loader module is used to program an application bitstream on FPGA.

4.2.1 Open-source VLSI tools

An open-source VLSI tool kit ALLIANCE [Alliance, 2006] and a python based language STRATUS [Belloeil et al., 2007] are used for the development of this FPGA layout generator. Alliance is a complete set of free CAD tools and portable CMOS libraries for VLSI design. It includes a VHDL compiler and a simulator, logic synthesis tools, and automatic place and route tools. STRATUS generates parameterized VLSI modules. It extends the python language with a set of methods and functions for procedural generation of netlist and layout views of structural cell based designs. One of the major benefits of using open-source VLSI

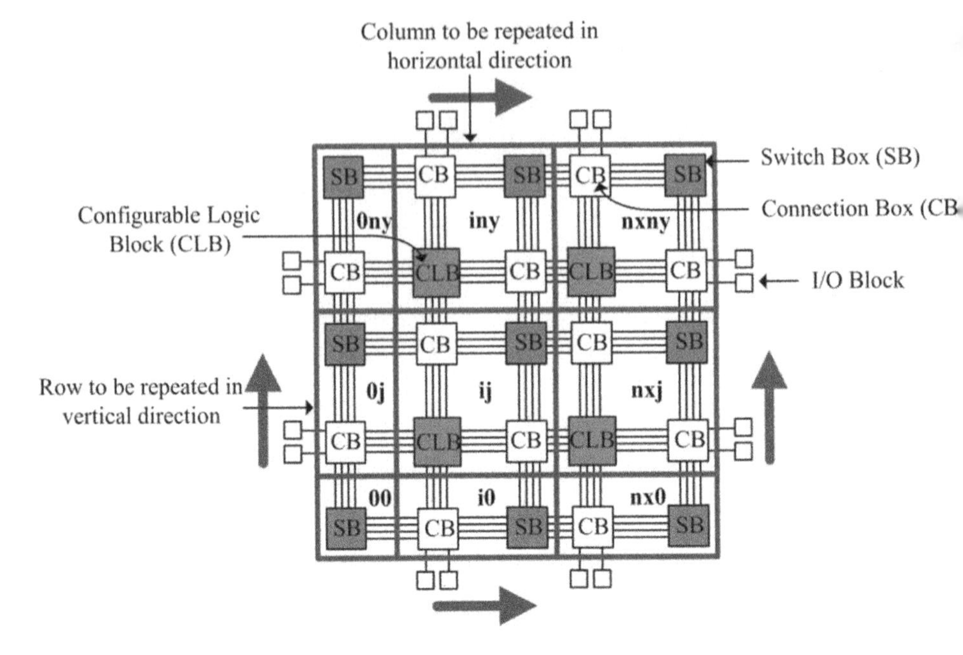

Figure 4.2: Tiles for Mesh-based FPGA

tools is that they can be easily adapted and modified for any specific demands.

4.2.2 Tile based approach

A tile based approach is used to generate the desired FPGA architecture. In this approach, a set of tiles are identified in the architecture which are repeatedly abutted to form the whole FPGA matrix. A set of 9 different tiles (as shown in Figure 4.2) are used for the generation of the target architecture. The principle tile is the tile 'ij', whereas other tiles are its derivations. The tiles on the extreme left column and the bottom row do not contain logic blocks. They only contain routing channel which connects adjacent I/O pads, and inputs of adjacent logic blocks. The rest of the tiles contain a top horizontal channel, a right vertical channel, a switch box, a logic block, and connection boxes. It can be seen in Figure 4.2 that the horizontal repetition of 2nd column and the vertical repetition of 2nd row generate an FPGA of our desired size. An important aspect in the tile based design is that the adjacent sides of two abutted tiles must have same length. While deciding the sizes of the tiles, priority is given to the tile which is the most area consuming; in this case the tile 'ij' is the most area consuming. The sizes of the other tiles are adjusted accordingly.

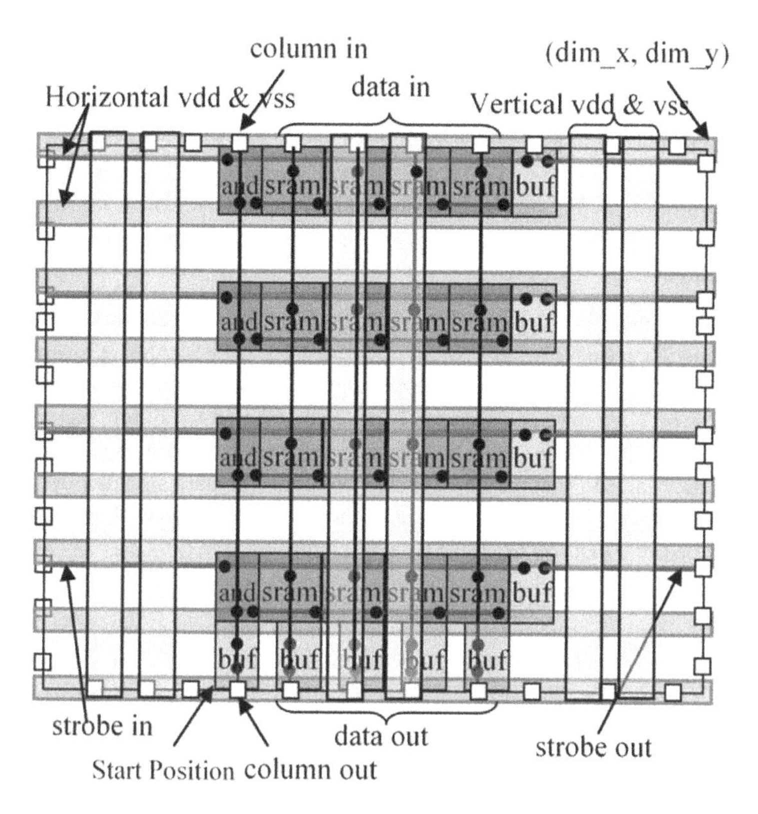

Figure 4.3: Partial layout of a sample FPGA Tile

4.2.3 Netlist generation

Each tile generator is written in STRATUS [Belloeil et al., 2007]. The tile generator receives a set of architectural parameters as input. It then generates the netlist of the tile in accordance with the given parameters. Loops and conditional statements are used to generate a tile for different parameters. The netlist of each tile is generated directly using the standard cell library named SXLIB. It is a symbolic standard cell library which comes with the ALLIANCE tool chain [Alliance, 2006]. C++ routines are also merged in the tile generator for generating VHDL model of specific components. These components are synthesized by the Alliance synthesizer named as BOOG. After synthesis, these components are used by the tile generator. The generated netlists of all the tiles are passed to the FPGA generator which links them together to construct the netlist of a complete FPGA. This generated FPGA netlist can be integrated as an embedded FPGA in larger applications.

4.2.4 Tile Layout

A tile generator generates both the netlist and the partial layout of a tile. The partial layout is generated with the help of parameterized algorithms which take a set of layout parameters as its input. Currently, the partial layout is performed for the generation of a regular memory configuration layout, proper buffering of few long wires, power routing, and a balanced clock distribution network. Later-on, the placer and the router are used to complete the remaining layout.

The partial layout generation algorithm places all the SRAM bits in rows and columns with a fixed distance between each row, as shown in Figure 4.2.4. Each SRAM bit in a row receives a vertical data signal, and a horizontal strobe signal. The data bits are written in all the SRAMs of a row only when strobe signal is high for that row, and column signal is high for the complete tile. The column and strobe signals come from bitstream configurator (loader), which is discussed later in section 4.4.2. The column and the data signals from the top are buffered before they exit on the bottom side of the tile. Similarly a strobe signals from the left is buffered before it exits on the right side of the tile.

The partial layout algorithm starts placing the bitstream configuration cells (SRAMs, buffers, AND gates etc) from a layout parameter named "Start Position" as shown in Figure 4.2.4. Similarly the height and width of a tile and the total SRAM bits in a tile are also variable parameters which change for different architectural features. For this purpose a small database is created which specify layout variables for different architectural parameters. The layout algorithm and the database specification are generic enough to handle various architectural parameters.

4.2.5 Power routing

The layout generation algorithm generates horizontal and vertical power segments as shown in Figure 4.2.4. The alternating VDD and GND segments in the horizontal direction are fixed whereas the placement of vertical power segments is supported by few layout parameters. The total number of vertical segments for power and ground in a tile, their positions and their widths are defined in the layout database. These values can be changed for tiles of different sizes. The horizontal power segments use the 1st and 2nd routing layer; whereas the vertical power segments use the 5th routing layer. The remaining partial tile layout is done using only the first 4 routing layers.

4.2.6 Clock generation

In this work, we have used a tile based approach for routing a symmetric H-tree clock distribution network. It is found that a group of 13 tiles can be used to generate a clock tree for a matrix of size 2^N x 2^N where N > 1. Each corresponding clock tile is automatically merged

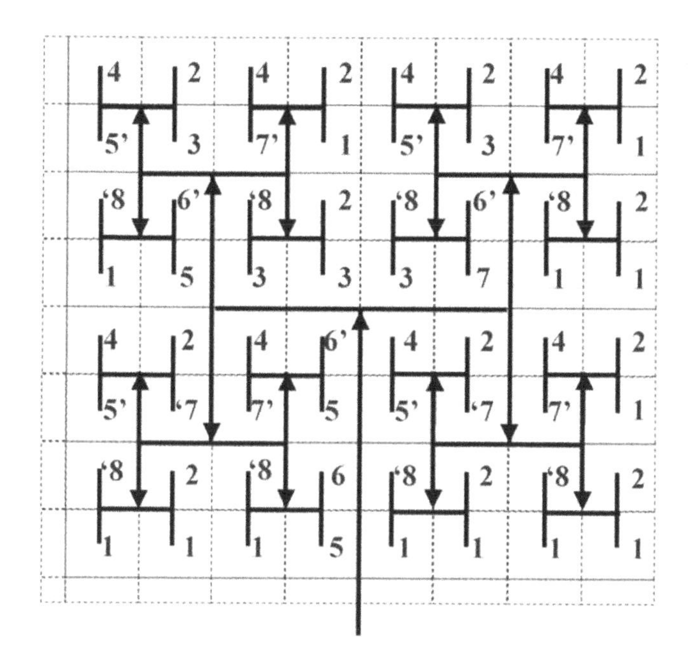

Figure 4.4: H-Tree clock distribution network for 8x8 FPGA

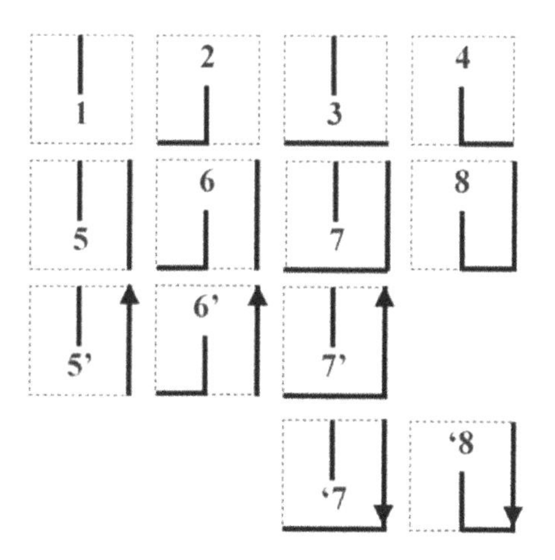

Figure 4.5: Tiles for construction a H-Tree clock distribution Network

with the FPGA tiles during the partial layout phase. This results in the generation of multiple copies of the same FPGA tiles having different clocks. After merging of FPGA and clock tiles, 23 different tiles are produced. These tiles can be abutted together to construct any FPGA of size 2^N x 2^N having a symmetric H-tree clock distribution network. A sample 8x8 clock distribution matrix is shown in Figure 4.4. Different clock tiles used in the generation of a clock tree are shown in Figure 4.5. An arrow in clock tiles represent buffers.

The main advantage of this mechanism is that we have a generic, tile-based balanced clock distribution network. One of the disadvantage is that it limits the FPGA size in X and Y direction to be equal and power of 2, i.e 4x4, 8x8, 32x32, 64x64 and so on. But since the clock generation algorithms and their merging with the FPGA tiles is totally automatic, generic algorithms can be written to route other types of clock distribution networks [Friedman, 2001]. The only thing to consider for writing a new clock routing algorithm is that placement of clock buffers and clock routing must not overlap with the partial layout. Currently the clock is routed in the 5th and 6th routing layer.

4.2.7 Pin generation

In a tile based FPGA, tiles connect together through abutment (i.e they are placed adjacent to each other). So the pin locations on the boundaries of adjacent tiles must overlap. The positions of few of these pins are calculated on the basis of the layout parameters found in the database. In Figure 4.2.4, the pins named "data in/out", "column in/out" and "strobe in/out" are dependent upon the layout parameters in the database. Since the database is common for all the tiles, so the pin abutment problem does not arise for these pins. There are some other pins which do not have a fixed position. These pins are automatically placed by placer. Since the final automatic placement of all the tiles is done independently, it is difficult for the placer to correctly choose the pin locations of the tiles. So a generic algorithm is written to choose overlapping positions of remaining pins in all the tiles. This algorithm places the pins in all the four boundaries of a tile and ensures that the pins are not congested to a limited area on the boundary of a tile. It utilizes all the available space and tries to distribute the pins with equal spacing.

4.2.8 Automatic placement and routing

After the partial layout generation of all the tiles, each tile is separately placed and routed with the help of ALLIANCE automatic placer and router named OCP and NERO respectively. Tile generation flow is shown in Figure 4.6(a). The partial layout information and the netlist of a tile is given to the placer to place the remaining logic of the netlist. If the placer is unable to place the design, the dimensions of the tile are manually increased in the database. The X and Y dimensions of the tile must be properly adjusted to make sure that a tile does not waste any extra space. The placer automatically adds the empty cells to fill up any extra space. After placement, NERO routes the whole design. All the tiles are successfully routed using 4 routing layers. Only the clock and the vertical power segments are routed on the 5th

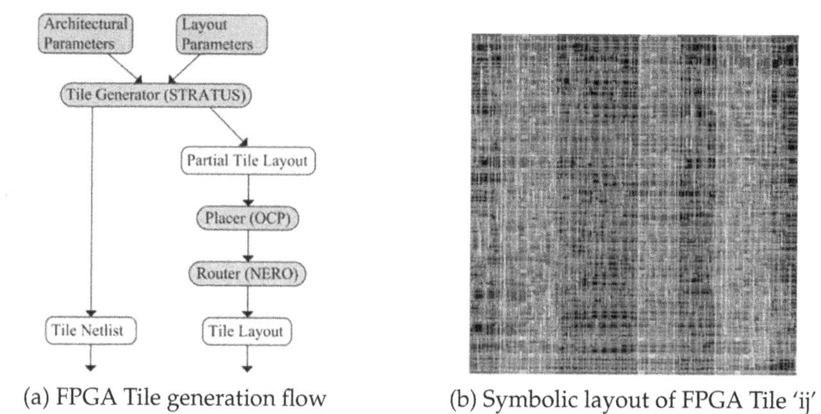

(a) FPGA Tile generation flow (b) Symbolic layout of FPGA Tile 'ij'

Figure 4.6: Symbolic layout generation of an FPGA Tile

and 6th routing layer. The complete layout of tile "ij" (after partial and automatic layout) is shown in Figure 4.6(b). The vertical yellow lines in the Figure represent the vertical power segments on 5th routing layer.

4.3 Architecture Features

The above process of FPGA generation is used to generate a bidirectional mesh-based FPGA with hardware support for the mitigation of Single Event Upsets (SEU) [Elder et al., 1988]. This section describes only a brief introduction of SEU mitigation; more details of the Redundant FPGA Core architecture supporting the detection and correction of SEUs can be found in [Mrabet, 2009]. SEU are induced by energized particles hitting the silicon device. A particle hit with sufficient energy changes the logic state of the memory elements, thus producing a transient error. An SEU on configuration bits may change the functionality of the Look-Up Tables as well as the interconnect controlled by SRAM cells. If the SRAM controlling a tristate based bidirectional interconnect is changed by an SEU, there is a possibility that two drivers write on the same wire, thus producing a hard error (which might damage the hardware). These hard errors can be eliminated by using simple decoders, as shown in Figure 4.7(a) and 4.7(b). The decoder system implements a dependency between switches that derive the same track. In this way, even if SEU changes an SRAM cell, no two drivers will be writing on the same wire. The hard error is avoided due to the introduction of decoders, however the functionality changes due to SEU. This change of functionality is called as a soft error. Soft error is a configuration error, that changes the functionality of a circuit, but does not harm the hardware. Hard error is not caused if single-driver unidirectional routing network is used. On the other hand, a soft error is generated by SEU in both bidirectional and unidirectional routing channels. An error detection system is integrated in each tile to detect the soft errors, as shown in Figure 4.7(c). The error detector implements parity system which detects a change in a single SRAM. The error detector for 8 bits is shown in Figure 4.7(d).

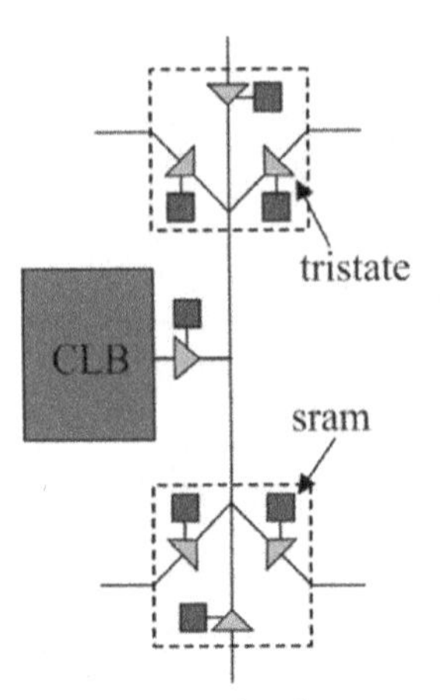

(a) Standard system for deriving a tristate based bidirectional routing track

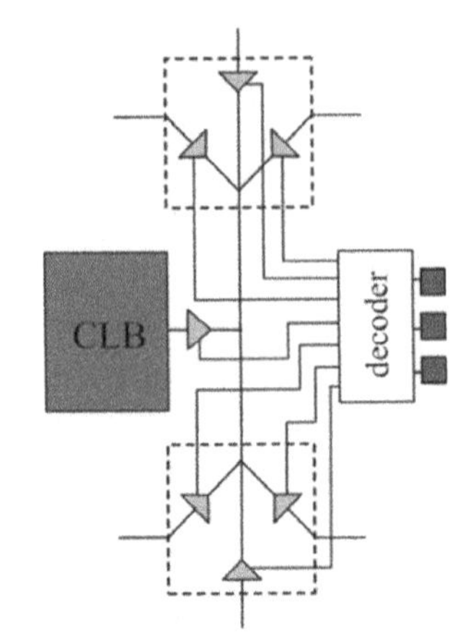

(b) Decoder system for deriving a tristate based bidirectional routing track

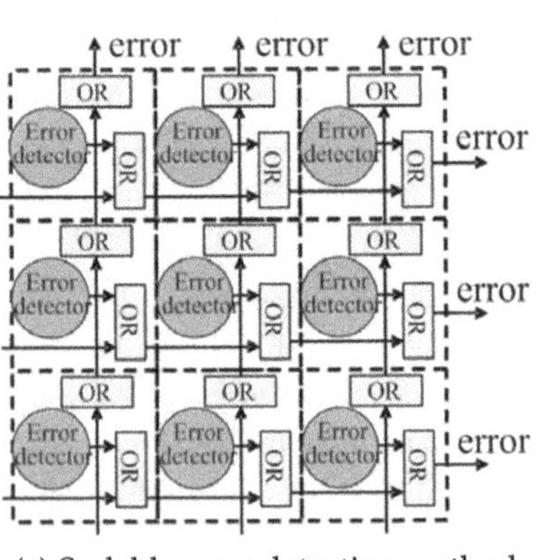

(c) Scalable error detection method

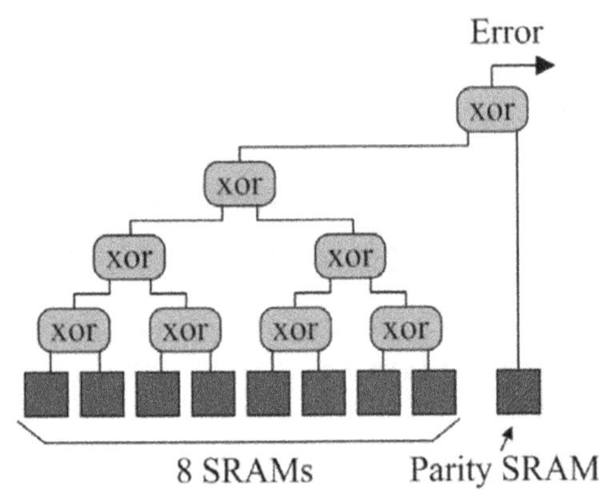

(d) Parity system of error detector

Figure 4.7: Error detection method to mitigate single event upset

The error detection system enables an error signal if a change is detected in configuration bits. The error signal propagates through row and column (as shown in Figure 4.7(c)). Once the error is detected, a particular tile or the group of tiles can be reconfigured to remove the error. According to final application requirements, a bidirectional 32x32 FPGA matrix with a channel width of 8 is generated and taped-out.

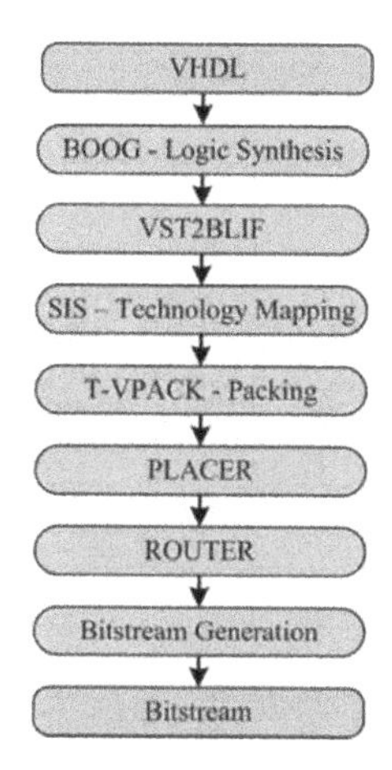

Figure 4.8: Software Flow

.4 Validation

4.1 Software flow

order to test the functionality of the generated architecture, a software flow is followed as own in Figure 4.8. The sample application (in VHDL format) that is to be mapped on the 'GA is the input to the software flow. Initially BOOG synthesizes the VHDL input into a etlist of gates (in VST format). VST2BLIF and later SIS is used to convert it into LUT-4 form. VPACK and later PLACER and ROUTER are used for the placement and routing of the etlist on the FPGA architecture. A bitstream generator is written which generates a binary ream that contains configuration information for configuring the sample application on the 'GA.

4.2 Bitstream configuration mechanism

tstream configuration mechanism for a sample 3x3 FPGA is shown in Figure 4.9. An Nx by y FPGA contains (Nx+1) by (Ny+1) tiles, where Nx+1 is the total number of tile columns d Ny+1 is the total number of tile rows in an FPGA. Each FPGA tile comprises a set of RAM bits arranged in multiple rows. This arrangement of SRAMs in rows is attained by e partial layout generator explained earlier. The SRAM bits in a row are called as a 'word'. row number, a column number and a word number is required to write a data word to tile. The row and column numbers give the location of the tile in a matrix, whereas the ord number gives the location of word in a tile. All these three parameters are passed to e shift registers through a single I/O Pad. The data to be written in a word is also specified the same shift registers. The row, column and word decoders are used to turn on the exact obe and column signals. Thus when write enable turns high, the data is written onto the

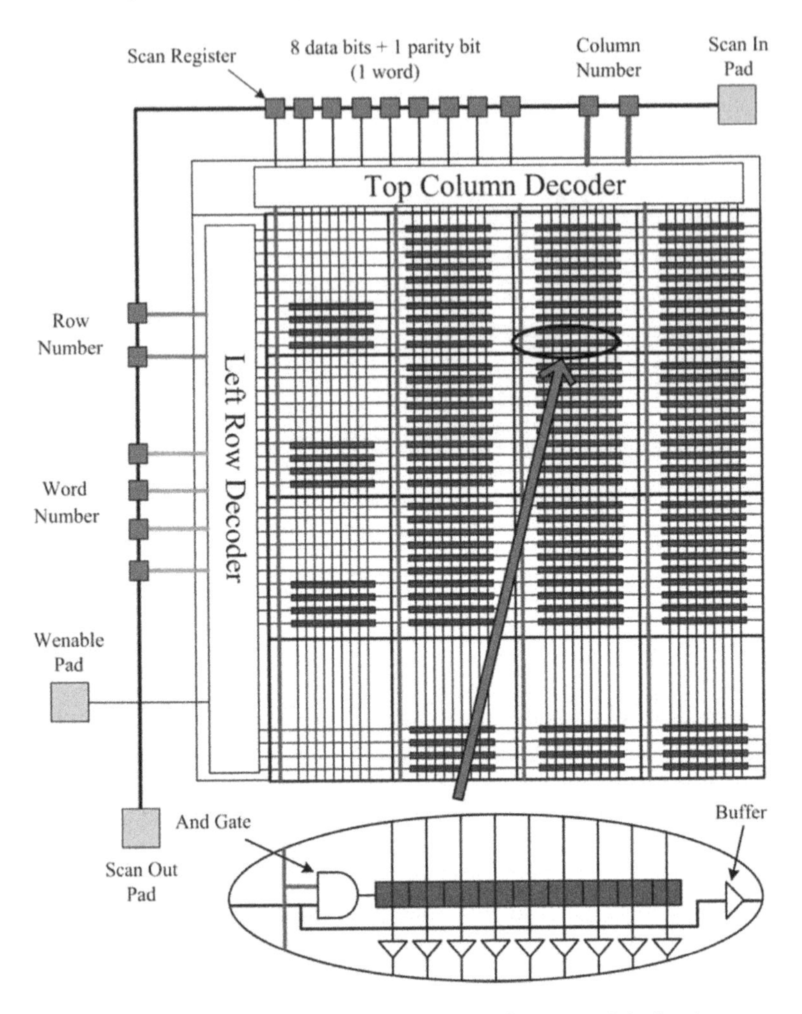

Figure 4.9: Bitstream Configuration Mechanism

specific word of the requested tile. This process is repeated for all the words of all the tile The shift registers and decoder are implemented in a loader which is also generated by th FPGA generator.

4.4.3 Simulation

The generated FPGA is tested on the ALLIANCE simulator called ASIMUT. Different te applications are tested on the FPGA. Each application is mapped on the FPGA with th help of the software flow described in section 4.4.1. The generated bitstream is loaded c the FPGA. Once the FPGA is programmed, respective testbench of each test application

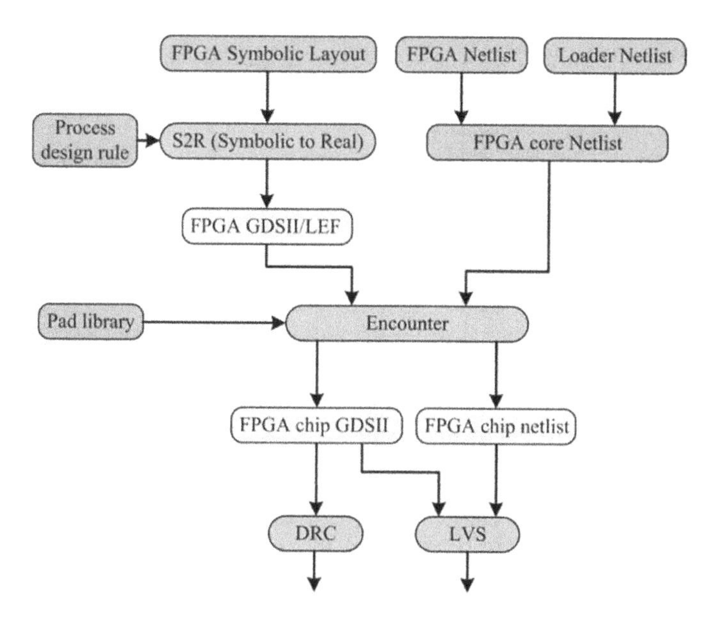

Figure 4.10: CAD flow for FPGA chip generation, verification and validation

applied on the inputs of the FPGA, and the outputs are compared. These simulations are also verified on other commercial tools like SYNOPSYS [Synopsys, 2010].

4.4.4 Netlist layout comparison

The generated netlist and the generated layout must match with each other. For this purpose the ALLIANCE extraction tool, COUGAR is used. It extracts a netlist from a layout. Later the ALLIANCE comparison tool, LVX, is used to compare the extracted netlist with the generated netlist. This confirms that the automatic layout matches with the automatic generated netlist. All the tiles are individually extracted and compared with their respective netlists. The flattened 32x32 FPGA matrix is too large and could not be compared due to the limitations of COUGAR.

4.4.5 Electric Simulation

The ALLIANCE extraction tool, COUGAR, is used to extract the spice model of each tile. These models are later electrically simulated using ELDO. Our extraction tool is unable to support very large circuits. So it was impossible to electrically simulate the complete 32x32 FPGA. However, for the proof of concept, electric model of a 4x4 FPGA matrix with channel width of 8 is successfully simulated.

4.5 Tapeout

A CAD flow for the tapeout, verification and validation of an FPGA chip is shown in Figure 4.10. The layout is generated by using symbolic standard cell library (SXLIB) which works on unit λ (lambda). The ALLIANCE tool S2R (symbolic to real) is used to convert the symbolic design to 130nm process technology.

The GDS and LEF files of the 32x32 FPGA layout are retrieved from S2R. The 32x32 FPGA occupies an area of 3885.6 μm by 3882 μm. It has been noted that 19% of the FPGA area increases due to the hardware support for the mitigation of SEU. Similarly 20% of area increases due to the generic symbolic design rules. The GDS and LEF files of an FPGA represent it as a black box, which can be used in some larger design. For the proof-of-concept, it is used to lay out a complete FPGA chip.

ENCOUNTER is used to lay out an FPGA chip. The chip comprises of a 32x32 FPGA matrix, loader, and IO pads of the chip. The symbolic layout of loader is not used; the layout of loader is regenerated by ENCOUNTER using 130nm standard cell library of STMicroelectronics [STMicroelectronics, 2010]. The height of cells in SXLIB and ST standard cell libraries are different, which eventually requires different power routing for loader. Thus, the power routing already performed in the FPGA matrix must not come in contact with the power routing done by ENCOUNTER for loader. One of the main reasons for not using the symbolic layout of loader is to test the integration the FPGA black box in a larger design which uses different standard cell library. ENCOUNTER modifies the loader netlist for proper clock distribution for scan registers in loader. However, ENCOUNTER does not change the netlist and layout of 32x32 FPGA matrix. The pads are placed and routed using ENCOUNTER. The DRC and LVS verification of the final chip is performed using CALIBRE. The final FPGA chip layout is shown in Figure 4.11, it measures 23.86 mm^2. The inner FPGA core (FPGA matrix + loader) measures 15.48mm^2.

4.6 Conclusion and Future Work

This chapter presented an automatic method for generating FPGA layout using an open-source VLSI tool-kit. The generator employs an elegant scheme to integrate manual intervention in the FPGA layout generation procedure. This is done by generating layout in two phases. In the first phase, a partial layout is generated using generic parameterized algorithms. The partial layout is generated to obtain a fast bitstream configuration mechanism an optimized power routing and a balanced clock distribution network. In the second phase the generator completes the remaining layout using automatic placer and router. The FPGA layout is generated using a symbolic standard-cell library which allows to migrate the symbolic layout to any fabrication process technology. The proposed method is validated by generating layout of a 1024 Look-Up Table based mesh FPGA architecture. The layout is successfully taped-out using 130nm 6-metal layer CMOS process of STMicroelectronics [STMicroelectronics, 2010].

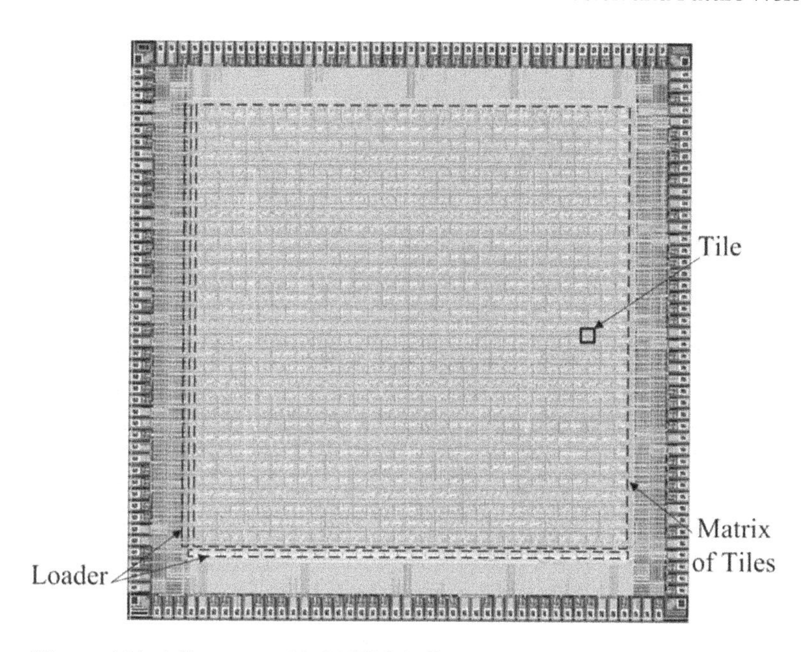

Figure 4.11: A Prototype 32x32 FPGA Chip Layout using 130 nm process

In future, this generator can be easily extended to support other architectural parameters such as unidirectional routing network, heterogeneous blocks and different clock distribution networks. The conversion from symbolic to real layout causes some area loss. The tile-based generator can be modified to directly generate the layout using real standard cell library.

5

ASIF: Application Specific Inflexible FPGA

An Application Specific Inflexible FPGA (ASIF) is an FPGA with reduced flexibility that can implement a set of application circuits which will operate at mutually exclusive times. These circuits are efficiently placed and routed on an FPGA to minimize total routing switches required by the architecture. Later all unused routing switches are removed from the FPGA to generate an ASIF. An ASIF for a set of 17 MCNC benchmark circuits is found to be 5.43 times (or 81.5%) smaller than a mesh-based unidirectional FPGA required to map any of these circuits.

5.1 Introduction and Previous Work

Low volume production of FPGA-based systems are quite effective and economical because they are easy to design and program in shortest possible time. The generic reconfigurable resources in an FPGA can be programmed to execute a large variety of applications. This flexibility of an FPGA enables implementation of different circuits on it at mutually exclusive times, or even allows dynamic reconfiguration for varying requirements. However, all these advantages come with a huge cost. FPGAs are much larger, slower, and more power consuming than their counterpart ASICs (Application Specific Integrated Circuits) [I.Kuon and J.Rose, 2007]. Consequently, FPGAs are unsuitable for applications requiring high volume production, high performance or low power consumption.

An ASIC has speed, power and area advantages over an FPGA, but at the expense of higher non-recurring engineering (NRE) cost and higher time-to-market. However, the NRE cost and time-to-market are reduced with the advent of a new breed of ASICs known as Structured-

H. Parvez and H. Mehrez, *Application-Specific Mesh-based Heterogeneous FPGA Architectures*,
DOI 10.1007/978-1-4419-7928-5_5, © Springer Science+Business Media, LLC 2011

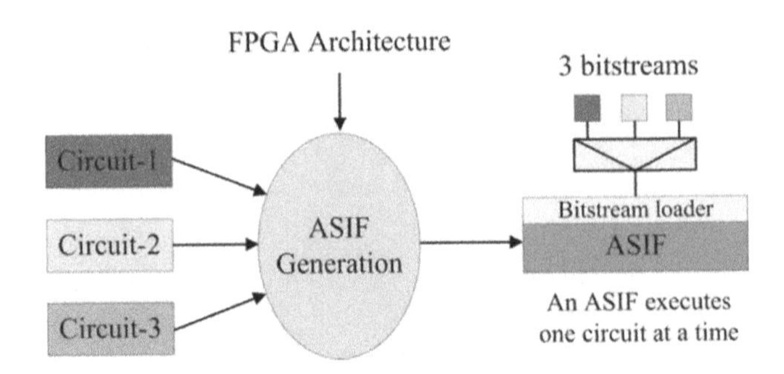

Figure 5.1: An illustration of ASIF generation concept

ASIC. Structured-ASICs contain array of optimized elements which implement a desired functionality by making changes to few upper mask layers. Structured-ASICs are explored or manufactured by several companies [Wu and Tsai, 2004] [Okamoto et al., 2004] [Sherlekar, 2004] [eASIC, 2010].

FPGA vendors have also started giving provision to migrate FPGA based application to Structured-ASIC. The main idea is to prototype, test and even ship initial few copies of a design on an FPGA; later it can be migrated to Structured-ASIC for high volume production [Hutton et al., 2006]. In this regard, Altera has proposed a clean migration methodology [Pistorius et al., 2007] that ensures equivalence verification between FPGA and its Structured-ASIC (known as HardCopy [HardCopy, IV]). However, migration of an FPGA based application to HardCopy can execute only a single circuit. HardCopy totally loses the quality of an FPGA to use the same hardware resources for executing multiple applications at different times. So, we propose a new reduced FPGA that can execute a set of circuits at different times. This reduced FPGA is termed as an Application Specific Inflexible FPGA (ASIF). Figure 5.1 illustrates the ASIF generation concept. When an FPGA-based product is in the final phase of its development cycle, and if the set of circuits to be mapped on the FPGA are known, it can be reduced for all the given set of circuits. An ASIF can yield considerable area and performance gains to an FPGA-based product by reducing it to a much smaller multiplexed circuit. Execution of different application circuits can be switched by loading their respective bitstream on ASIF.

The concept of an ASIF is similar to configurable ASIC cores, abbreviated as cASIC [Compton and Hauck, 2007]. cASIC is a reconfigurable device that can implement a given set of circuits which operate at mutually exclusive times. However, cASIC and ASIF have several major differences. cASIC is intended as an accelerator in a domain-specific systems on-a-chip, and is not designed to replace an entire ASIC-only chip. For that reason, cASIC supports only full-word logic blocks (such as 16-bit wide multipliers, adders, RAMs etc to implement data-path circuits. However, an ASIF supports both fine- and coarse-grained logic blocks. The routing network used by cASIC and ASIF are totally different. cASIC uses 1-D segmented bus interconnect, whereas ASIF uses 2-D mesh interconnect. Another major difference between cASIC and ASIF is the approach with which their routing network re optimized. cASIC is generated using a constructive bottom-up "insertion" approach; re-onfigurability is inserted through the addition of multiplexers and demultiplexers. On the ontrary, an ASIF is generated using an iterative top-down "removal" technique; different ircuits are mapped onto an FPGA, and flexibility is removed from the FPGA to support only ne given set of circuits. The benefit of a "removal" approach over an "insertion" approach that any existing FPGA architecture can be reduced to an ASIF using this "removal" tech-ique.

his chapter mainly concentrates on different ASIF generation techniques, experimentation nd analysis of an ASIF using homogeneous logic blocks, and quality analysis of ASIF gen-ration techniques. Chapter 6 will present ASIFs using heterogeneous logic blocks, and how exploit their limited re-programmability to map new applications on ASIFs. This work onsiders only area optimization of ASIFs, delay and power parameters are not explored.

.2 Reference FPGA architecture

his section describes a reference FPGA architecture used in this work. Different netlists e mapped on this architecture that is later reduced to an ASIF. The reference FPGA is a esh-based VPR-style (Versatile Place & Route) [Betz et al., 1999] architecture, as shown in gure 2.1. It contains Configurable Logic Blocks (CLBs) arranged on a two dimensional grid. ach CLB contains one Look-Up Table with 4 inputs and 1 output (LUT-4), and one Flip-op (FF). A CLB is surrounded by a single-driver unidirectional routing network [Lemieux al., 2004]. FPGA is divided into "tiles" that are repeated horizontally and vertically to rm a complete FPGA. A single FPGA tile, surrounded by its neighbouring tiles is shown Figure 3.2. Each of the 4 inputs of a CLB are connected to 4 adjacent routing channels. he output pin of a CLB connects with the routing channel on its top and right through the agonal connections of the switch box (highlighted in the bottom-left switch box shown in gure 3.2). This unidirectional disjoint switch box connects uniform length routing tracks r wires). The connectivity of routing channel with the input and output pins of a CLB, bbreviated as Fc(in) and Fc(out), is set to be maximum at 1.0. The channel width is varied cording to the netlist requirement but remains in multiples of 2 [Lemieux et al., 2004].

circuit or netlist to be mapped on an FPGA is initially transformed in the form of CLBs UTs and/or FF). A netlist file (in .NET format) includes I/O (Inputs and Outputs), CLBs UTs and/or FF) instances which are interconnected through signals called NETS. A soft-are module named PLACER uses the simulated annealing algorithm [Betz et al., 1999] [Kirk-trick et al., 1983] to place the IO/CLBs instances on their respective blocks of FPGA. The unding box (BBX) of a NET is a minimum rectangle that contains the driver instance and receiving instances of a NET. The PLACER attempts to achieve a placement with mini-um sum of half-perimeters of the bounding boxes of all NETS. It moves an instance ran-omly from one block position to another. After each move operation, the BBX cost is up-ted incrementally. Depending on cost value and annealing temperature the operation is cepted or rejected. After placement, a software module named ROUTER routes the netlist

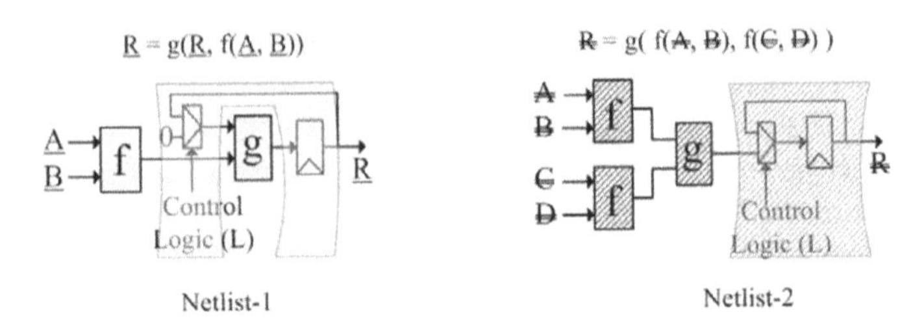

Figure 5.2: Example Circuits

on the architecture. The router uses a pathfinder algorithm [L.McMurchie and C.Ebeling, 1995] to route netlists using FPGA routing resources.

5.3 ASIF Generation Techniques

This section discusses various techniques for the generation of an ASIF. An ASIF is an FPGA with reduced flexibility that can implement a set of circuits exclusively. It is generated by removing unused configurable routing resources of an FPGA. This section presents different techniques of mapping netlists on an FPGA so that final ASIF occupies least possible area. An ASIF is generated by initially defining a minimum FPGA architecture that can map any netlist belonging to the given set of netlists. Next, the given netlists are efficiently placed and routed on the FPGA to minimize the total routing switches required by the architecture. Finally all the routing switches that are not used by any of the given set of netlists are removed from the FPGA to generate an ASIF. When switches are removed from FPGA, long wires are created. Buffers are added to cut these long wires. It is worth noticing here that once an ASIF is generated, the placement and routing of the given set of netlists is fixed. Even a slightest change in the placement or routing of a netlist is not guaranteed, but not impossible either.

An ASIF can be generated for any set of circuits that we happen to need for a system; the circuits must not necessarily belong to similar application domain. In this section, different ASIF generation methodologies are explained with the help of two example circuits shown in Figure 5.2. Later in section 5.6, the same techniques will be applied on a larger set of netlists. Initially, a target FPGA architecture is generated with maximum number of CLBs required by the given set of netlists. If the number of CLBs in an FPGA does not make a perfect square, FPGA size is selected to be a nearest rectangle. Both the netlists are then routed on the target architecture with minimum channel width. Figure 5.3(a) and 5.3(b) show the two netlists placed and routed individually on the target FPGA. Wires used by netlist-1 are continuous/blue, whereas wires used by netlist-2 are dotted/black. Grey wires are the unused wires in an FPGA. For simplification, Figure 5.3 does not show the inner connection details of switch box and connection box.

Four different ASIF generation techniques are discussed as below.

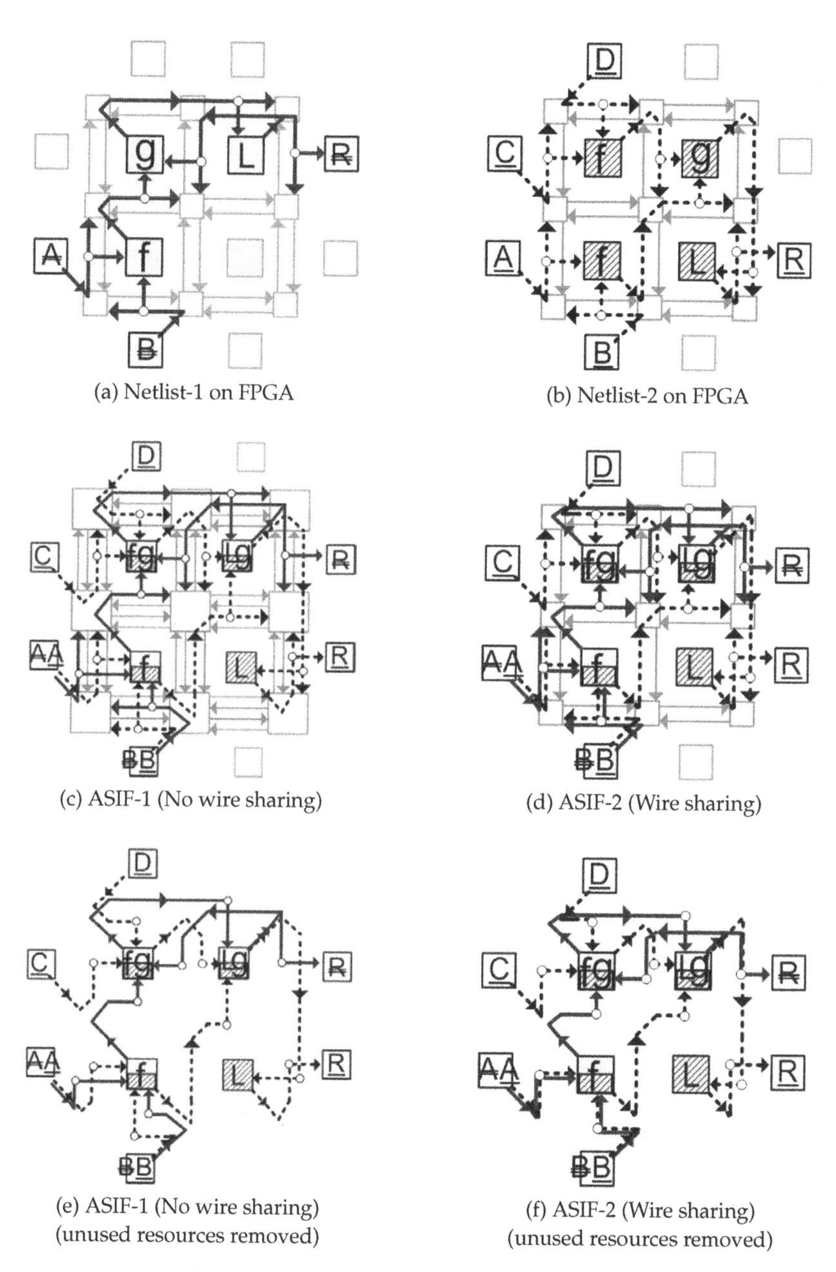

(a) Netlist-1 on FPGA

(b) Netlist-2 on FPGA

(c) ASIF-1 (No wire sharing)

(d) ASIF-2 (Wire sharing)

(e) ASIF-1 (No wire sharing) (unused resources removed)

(f) ASIF-2 (Wire sharing) (unused resources removed)

Figure 5.3: Example circuits mapped on FPGA

5.3.1 ASIF-1 (No wire sharing)

After selection of FPGA size, all netlists are placed separately on the FPGA. The CLB/IO instances of different netlists share the CLB/IO blocks on the architecture. But routing of all netlists is done without sharing routing wires over all netlists. Each netlist uses its own minimum routing channel network. Thus the target FPGA architecture contains maximum number of CLBs required by any netlist, whereas the channel width of the target FPGA is the sum of the channel widths required by all netlists.

Figure 5.3(c) shows the FPGA architecture on which both example netlists are mapped with no wire sharing. It has a channel width of 4. Two routing channels are used by netlist-1 and the other two are used by netlist-2. Finally all unused switches in the switch box and the connection box are removed to generate an ASIF. Figure 5.3(e) shows the ASIF after the removal of unused resources. Since no routing wires are shared between any of the netlists, ASIF-1 does not require any wire multiplexing in the switch box. Only the required switches of the connection box are retained. The routing of netlists without sharing routing wires can generate an ASIF with least number of switches. However, a major disadvantage of this technique is that the total number of switches in the connection box, and the total routing wires increase as the number of netlists increases. Consequently, the layout area of ASIF-1 for large number of netlists might eventually be dominated by routing wires.

5.3.2 ASIF-2 (Wire sharing)

In this method, the placement and routing of netlists is done in the same way as done normally for FPGAs. The netlists are placed and routed individually on the FPGA. The CLB/IO blocks on the architecture, and the routing channel are shared by all netlists. So the target FPGA architecture contains maximum number of CLBs and maximum channel width required by any of the netlist in the group of netlists. Figure 5.3(d) shows the FPGA architecture on which both example netlists are mapped with wire sharing. Figure 5.3(f) shows this ASIF after the removal of unused resources.

After the placement and routing of netlists, unused switches are removed. The retained switches in an ASIF can belong to a switch box or a connection box. The total wires required in this ASIF are relatively less when compared to the previous method. However, the number of switches in this ASIF can increase considerably, as routing of each netlist is done without considering the routing paths of other netlists.

5.3.3 ASIF-3 (Efficient wire sharing)

The main motivation behind this method is to combine the benefits of the previous two methods, i.e. less wires and less switches. In this method, the netlists are placed separately on the FPGA. But routing is done efficiently in order to minimize the required number of switches and routing wires. This is done by maximizing the shared switches required for

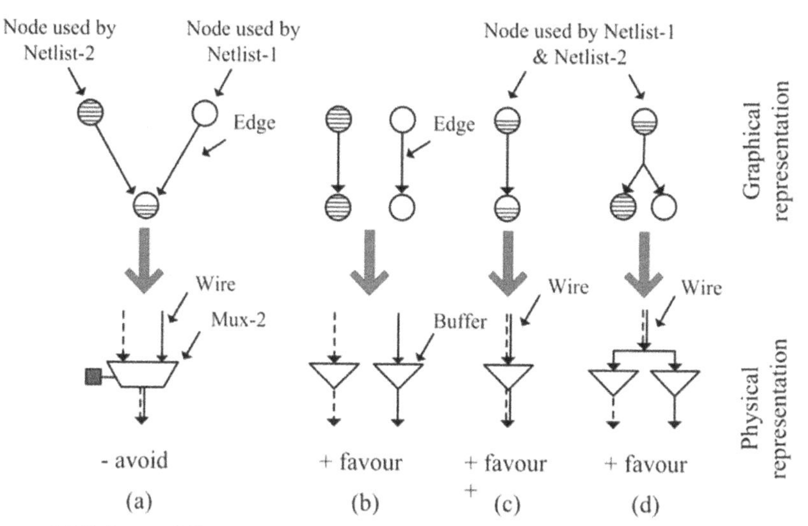

(a) Nodes use different edges to drive same node (b) Nodes use different edges to drive different nodes (c) Node uses the same edge to drive same node (d) Node uses different edges to drive different nodes

Figure 5.4: Efficient Wire Sharing

outing all the netlists on the FPGA. The efficient wire sharing encourages different netlists to oute their nets on an FPGA with maximum common routing paths. After all the netlists are fficiently routed on FPGA, unused switches are removed from the architecture to generate n ASIF.

he pathfinder routing algorithm is modified to support efficient wire sharing. Before we escribe these changes in detail, a short summary of the routing algorithm is presented ere. An FPGA routing network is represented by a graph with nodes connecting each other hrough edges; each routing wire of the architecture is represented by a node, and connecion between two wires is represented by an edge. When a netlist is routed on the FPGA outing graph, each net (i.e. connection of a driver instance with its receivers) is routed using a congestion driven "Shortest Path" algorithm. When all nets in a netlist are routed, one outing iteration is said to be completed. At the end of an iteration, there can be conflicts etween different nets sharing the same node; therefore congestion parameters are updated nd iteration is repeated until routing converges to a feasible solution (i.e. no conflicts are ound) or routing fails (i.e. maximum iteration count has reached, and few conflicts remain nresolved). Multiple netlists can be routed on the FPGA by allowing nodes to be shared by nultiple nets belonging to different netlists.

he pathfinder routing algorithm is modified to perform efficient wire sharing for the group f netlists. Netlists are routed in a sequence. Each routed netlist saves information that which odes and edges it has used. Later, the next netlist uses this information to perform efficient routing. Figure 5.4 explains different routing scenarios when 2 netlists are routed on a routing graph. The graphical representation in Figure 5.4(a) shows a case in which two nodes occupied by nets of 2 different netlists drive the same node. Figure 5.4(b) shows a case in which nodes occupied by nets of different netlists use different edges to drive different nodes. In Figure 5.4(c), both netlists share same node and edge to drive same node. Finally, Figure 5.4(d) shows a node shared by both netlists drives different nodes. The physical representation for each of the four sub graphs are also presented along. In order to reduce the total number of switches and total wire requirement, the physical representation in Figure 5.4 suggests that case (a) must be avoided because it increases the number of switches (here a mux-2), whereas case (b), (c) and (d) should be favoured. Favouring these cases means that if more routing resources exist in FPGA architecture, it is more probable to exploit such cases. For this reason, in order to create more routing resources, section 5.6 performs experiments with varying channel widths.

The routing preferences shown in Figure 5.4 need to be integrated in the pathfinder routing algorithm. For this purpose, the cost function of a node is modified in a similar way as for timing-driven cost function [Betz et al., 1999]. A particular routing is avoided or favoured by increasing or decreasing the cost of a node. If a net is to be routed from current node to next node, the cost of next node can be calculated with the formulas shown in Equation 5.1. The cost of a node depends on its congestion cost. Here, the increase or decrease in cost is controlled by a constant "Factor". The value of this factor ranges between 0 and 0.99. If an FPGA architecture has limited routing resources, a maximum value of factor might not allow the routing algorithm to resolve all congestion problems. So the value of factor is gradually decreased if the routing solution does not converge after a few routing iterations.

$$\begin{array}{ll} \text{(Normal)} & \text{Cost}(n) = \text{CongestionCost}(n) \\ \text{(Avoid)} & \text{Cost}(n) = (1 + \text{Factor}) * \text{CongestionCost}(n) \\ \text{(Prefer)} & \text{Cost}(n) = (1 - \text{Factor}) * \text{CongestionCost}(n) \\ & \text{Where } 0 \leq \text{Factor} \leq 0.99 \end{array} \tag{5.1}$$

Multiple netlists can be routed on an FPGA architecture either sequentially or parallelly. Sequential routing of netlists in a particular order is called here as "Netlist by Netlist" routing. In a "Netlist by Netlist" routing, each routed netlist saves the information of the routing path it has used. Later, the next netlist uses this information to perform efficient routing by giving preference to some nodes over others. Experiments are done with netlists sequenced in different orderings (i.e. netlists ordered in ascending or descending order according to their size, channel width, wire utilization and few random orders). An ASIF generated with netlists routed sequentially in descending order of their channel widths or of the number of routing wires used gives minimum area results. However the area difference between ASIFs (generated using different netlist orderings) becomes negligible as the routing resources increase.

In order to get rid of dependence to route netlists in a particular order, the netlists are routed parallelly. Two parallel techniques are tried; routing-iterations of different netlists are routed n a sequence (“Iteration by Iteration”) and nets of different netlists are routed in a sequence “Net by Net”). However, both techniques give much worse results than the best “Netlist by Netlist” ordering. This is because in parallel routing techniques, the routing of all the netlists emains incomplete simultaneously. In order to avoid congestion, the nets keep on changing heir paths in different routing iterations. A path that is chosen by netlist-2 because it was sed by netlist-1, might not eventually be used by netlist-1, but still remains in use by netlist-. Thus, due to inaccurate routing information, both parallel methods end up taking more witches than the best “Netlist by Netlist” results.

All experimental results shown in section 5.6 use the maximum value of “Factor”. The maxi-num routing iteration is set to 30, and the value of “Factor” is gradually decreased if routing oes not converge within the first 15 routing iterations.

.3.4 ASIF-4 (Efficient placement)

he placement of netlists can also be appropriately modified to get maximum benefit from fficient wire sharing. In all previous ASIF generation techniques, simulated annealing al-orithm with bounding box (BBX) cost function is used to place each netlist individually. In ther words, only intra-netlist placement optimization is performed. Intra-netlist placement eeps the placement of one netlist totally unaware of the placement of other netlists. Effi-ient placement is an inter-netlist placement optimization technique which can reduce the otal number of switches required in an ASIF. It tries to place driver instances of different etlists on a common block position, and their receiver instances on another common block. ater, efficient routing increases the probability to connect the driver and receiver instances f these netlists by using the same routing wires. This placement technique can be under-ood with an example shown in Figure 5.5. Figure 5.5(a) shows two very simple netlists; oth have the minimum possible BBX placement cost. The ASIF for these two netlists re-uires 4 multiplexors (mux-2). Figure 5.5(b) shows the same two netlists that are placed ficiently having same BBX cost as for netlists in Figure 5.5(a); the ASIF for these netlists equires no switches at all.

n order to perform efficient placement, a new cost function is proposed, named here as the Driver Count” (DC) cost function. This cost function calculates the sum of driver blocks tar-eting the receiver blocks of the architecture over all netlists. Efficient placement considers set of netlists simultaneously and aims at optimizing both intra-netlist and inter-netlist in-ance placements. The aim is to minimize the BBX cost of each netlist and the DC cost over l the netlists. In Figure 5.5(a) the ASIF has a “Driver Count” cost of 8, where each block as 2 different drivers. In Figure 5.5(b) the “Driver Count” cost is 4, with each block having nly one driver. Efficient placement uses both cost functions in parallel (i.e. the bounding ox (BBX) cost function and the Driver Count (DC) cost function). Since the BBX cost and e DC cost are not of same magnitude, initially both costs are made comparable by mul-plying one of them with a normalization factor. This factor is determined from the initial BX and DC costs. Weighting coefficients are attributed to them and a new weighted cost computed as shown in Equation 5.2. The simulated annealing algorithm later uses this

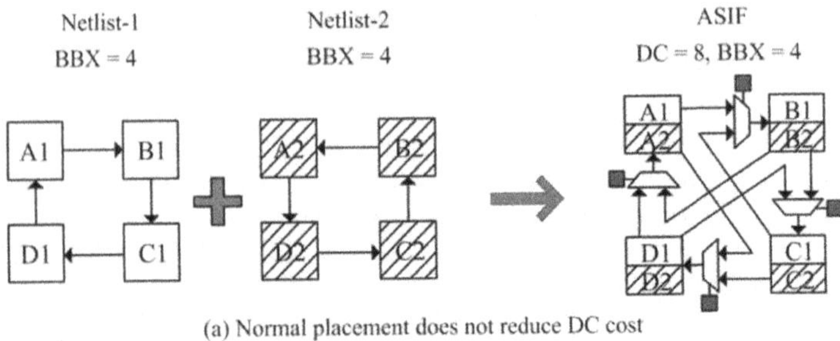

(a) Normal placement does not reduce DC cost

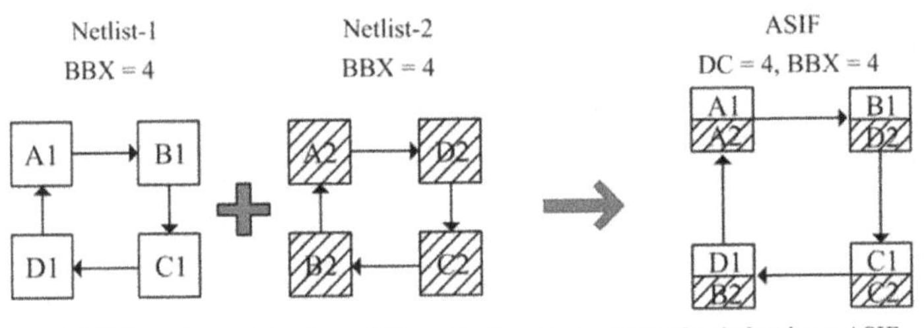

(b) Efficient placement reduces DC cost to decrease number of switches in an ASIF
Where BBX = Bounding box, DC = Driver Count

Figure 5.5: Efficient Placement

new weighted cost. It should be noted that as the weightage for DC function is increased the DC cost decreases, but the BBX cost increases, and vice-versa. With increase in the BBX cost more routing switches are required to route a netlist, which in turn means that more area is required. A compromise needs to be searched to obtain a good solution.

$$\begin{gathered}\text{Cost} = ((W * \text{BBX}) + ((100 - W) * \text{DC} * \text{NormalizationFactor})) \,/\, 100 \\ \text{Where } 0 \leq W \leq 100,\ \text{NormalizationFactor} = \text{Initial BBX} \,/\, \text{Initial DC} \\ \text{BBX} = \text{Bounding Box Cost},\ \text{DC} = \text{Driver Count Cost}\end{gathered} \tag{5.2}$$

Placement of multiple netlists is supported in the same way as used by cASIC [Compton and Hauck, 2007]. With multiple netlist placements, each block of the architecture can allow mapping of multiple instances belonging to different netlists. The placer is also modified to support the "Driver Count" cost function. All netlists are simultaneously placed. The placer randomly chooses an instance from any input netlist, and changes its position. The differences in the BBX Cost and the DC cost are updated incrementally. New weighted cost is calculated in accordance to the given weights; the simulated annealing algorithm uses this cost to decide if the movement is accepted or rejected. After all netlists are placed, efficient routing is performed. All unused switches are then removed to generate an ASIF.

5.4 ASIF Exploration Environment

The architecture exploration environment, presented in Chapter 3, is upgraded to explore different ASIF generation techniques. Following major additions are done in the exploration environment:

- **Support for routing multiple netlists**

  Efficient wire sharing encourages different netlists to share maximum common routing paths. Efficient wire sharing routes multiple netlists simultaneously, either sequentially or parallelly (as described in section 5.3.3). In both type of routings, the routing paths used by one netlist are preferred by other netlists. For this purpose, necessary changes are done in the exploration environment to support routing of multiple netlists. Different data structures are enhanced to maintain routing information of multiple netlists simultaneously. Support of placing multiple netlists simultaneously is already explained in section 3.4.2.

- **Routing congestion cost**

  Efficient wire sharing is implemented by giving preference to few nodes over others. The new formula for the calculation of node cost, as presented in Equation 5.1, is implemented in the congestion driven PathFinder routing algorithm. The avoid or preference factors in the Equation are defined in the architecture description file.

- **Driver count cost function**

  The driver count cost function is integrated in the simulated annealing based placement algorithm. The driver count cost is computed in parallel with the computation of bounding box (BBX) cost. A weighted average of both costs, as shown in Equation 5.2, is implemented to obtain a compromised placement solution. The driver count cost function simply calculates the sum of driver blocks targeting receiver blocks of the architecture for all the netlists. Since, the simulated annealing placement algorithm performs a lot of movements in each temperature step, the total execution time of the driver count cost function is reduced by implementing an incremental cost function. After each move operation of the placer, the incremental driver count function updates the driver count of only the effected blocks. The weightage parameters are defined in the architecture description file.

- **Support for no wire sharing**

  Multiple netlists are restricted to route on the architecture without sharing any routing wires with other netlists. This option is provided to explore an ASIF generation technique with no wire sharing (as described in section 5.3.1).

- **Support for reading pre-placed netlists**

  The architecture exploration environment is modified to support pre-placed netlists. The major motivation to support pre-placed netlists is to reduce execution time. When a netlist is placed on the FPGA architecture, the placement information of the netlist is stored in a placement file. The placement file contains position of netlist instances on the FPGA architecture. These placement files can be read by the exploration environment to avoid re-placing the netlists for some other experiments. After reading a placement file, all necessary data structures are initialized in the environment.

For ASIF generation, multiple netlists are placed simultaneously to perform inter-netlist placement optimizations. However, if large number of netlists are placed simultaneously, inter-netlist placement can take huge amount of time. Placement time can be reduced by placing only few netlists simultaneously. The placement generated for these few netlists is individually stored in their respective placement files. The remaining netlists can be efficiently placed along with the previously placed netlists. The placer does not change the instance positions of pre-placed netlists. Only the placement of non-placed netlists is optimized, keeping in consideration the placement of pre-placed netlists. In this way, considerable amount of time is saved, without any major compromise on the quality of the placement solution.

- **Support for reading pre-routed netlists**

  The architecture exploration environment is modified to support pre-routed netlists. When a netlist is routed on the FPGA architecture, the routing information of the netlist is stored in a routing file. The routing file contains the exact routing path of all the nets of a netlist routed on the FPGA architecture. This routing file can be read by the exploration environment to retrieve the exact routing path of all the nets of a given netlist. After reading the routing file, all necessary routing data structures are initiated in the environment.

  For ASIF generation, multiple netlists are efficiently routed together to maximize common routing paths. Any number of pre-routed netlist files can be read by the exploration environment to initialize necessary data-structures. The remaining netlists are efficiently routed, keeping in consideration the previously routed netlists. The pre-routed netlists can also be used to construct the routing network of the architecture Rather than constructing the routing network in a pre-determined manner (as described in a architecture description file), only those routing switches are constructed which are used by any of the pre-routed netlist. In this way, an ASIF can be re-construct by simply reading all the pre-placed and pre-routed netlist files.

- **Support for mapping new netlists on ASIF**

  A new placement algorithm is added to map new or modified netlist on an ASIF. The detailed description of this algorithm is presented in Chapter 6.

5.5 Area Model

A generic area model is used to calculate the area taken by the FPGA and different ASIFs The area model is based on the reference FPGA architecture shown in Figure 3.2. Area o

Index	Netlist Name	Number of CLBs	FPGA Size	Min Channel Width*
1.	pdc	4575	68x68	18
2.	ex5p	1064	33x33	16
3.	spla	3690	61x61	14
4.	apex4	1262	36x36	14
5.	ex1010	4598	68x68	12
6.	frisc	3556	60x60	12
7.	apex2	1878	44x44	12
8.	seq	1750	42x42	12
9.	misex3	1397	38x38	12
10.	elliptic	3604	61x61	10
11.	alu4	1522	40x40	10
12.	des	1591	40x40	8
13.	s298	1931	44x44	8
14.	bigkey	1707	42x42	8
15.	diffeq	1497	39x39	8
16.	dsip	1370	38x38	6
17.	tseng	1047	33x33	6
18.	clma	8383	92x92	12
19.	s38584.1	6447	81x81	10
20.	s38417	6406	81x81	8

* Minimum routing channel width on an FPGA sized 68x68 (except last 3 netlists)

Table 5.1: MCNC benchmark circuits

RAMs, multiplexers, buffers and Flip-Flops is taken from a symbolic standard cell library SXLIB [Alliance, 2006]) which works on unit Lambda(λ). When an ASIF is generated, all nused resources are removed. With the removal of switches, wires are connected with one nother to form long wires. A buffer is added for every wire of length 8 (spanning 8 CLBs) nd for every 8 wires driven by an output wire of a block. The area of FPGA or ASIF is eported as the sum of the areas taken by the switch box, connection box, buffers and CLBs. he area model also reports the total number of routing wires used for routing all netlists. In e next section the term "Routing area" is used for area taken by switch box, connection box nd buffers. The term "Logic area" is used for area taken by "CLBs". Since an ASIF reduces nly the routing area whereas the logic area remains constant; few of the area comparisons n the next section are done only for "Routing area".

5.6 Experimentation and Analysis

5.6.1 MCNC benchmark circuits

In this section, ASIF generation techniques are applied on a set of MCNC designs (Microelectronics Center of North Carolina designs) [S.Yang, 1991]. These benchmark circuits are shown in Table 5.1 in descending order of channel width requirement. Various experiments are done on the first 17 netlists. The last 3 netlists are not included in the experimentation because the FPGA sizes of these netlists are much larger and the channel widths much smaller than those of some other benchmarks in the list. Their inclusion increases the overall area of the target FPGA thus giving unnecessary area advantage to an ASIF. However for the sake of record, area results of an ASIF for all the 20 MCNC circuits are also presented at end of this section.

5.6.2 Results and Analysis

This section generates different ASIFs for a set of MCNC benchmarks, and then compares them with an FPGA. For the sake of clarity, experimental results for ASIFs are presented in different orderings than the order in which different ASIF generation methods are presented in section 5.3.

Figure 5.6 shows ASIF-4 routing area for 17 benchmark circuits. The X-axis shows variations in Bounding Box/Driver Count weighting coefficients. The Y-axis shows the routing area. The results are shown for different channel widths (between 18 and 192). It can be seen that routing area decreases as the channel width increases. This decrease in area is due to the efficient wire sharing. As more routing resources are made available, more there is a chance to find common routing path as represented in Figure 5.4(c), or more there is a chance to find non-congested independent routing paths as represented in Figure 5.4(b) & (d); eventually total area decreases. The second important thing to note is that the least routing area is found with BBX/DC weightage coefficient ratio of 80/20. For channel width 18 and 22, the BBX/DC ratio of 85/15 gives minimum area i.e 5.3% and 6.3% better as compared to BBX/DC ratio of 100/0. For other channel widths 80/20 gives the best results. The percentage gains for 80/20 as compared to 100/0 for channel widths 32, 48, 64, 96 and 192 are 8.6%, 9.4%, 9.6%, 8.0% and 8.2% respectively. The BBX/DC ratio less than 80/20 deteriorates the BBX cost which requires more routing resources, and thus area increases. For channel width 18, the BBX/DC ratio less than 70/30 deteriorates the BBX cost to such an extent that it becomes unroutable for channel width 18. This ratio of 80/20 is used for ASIF-4 in all other experiments.

Figure 5.7 shows routing area for different ASIFs with varying channel widths. ASIF-1 shares no routing wires between the netlists. Thus the total channel width required by ASIF-1 is the sum of channel widths required by 17 MCNC netlists. Due to this reason, the minimum channel width for ASIF-1 starts with 186 (i.e. sum of the minimum channel widths required

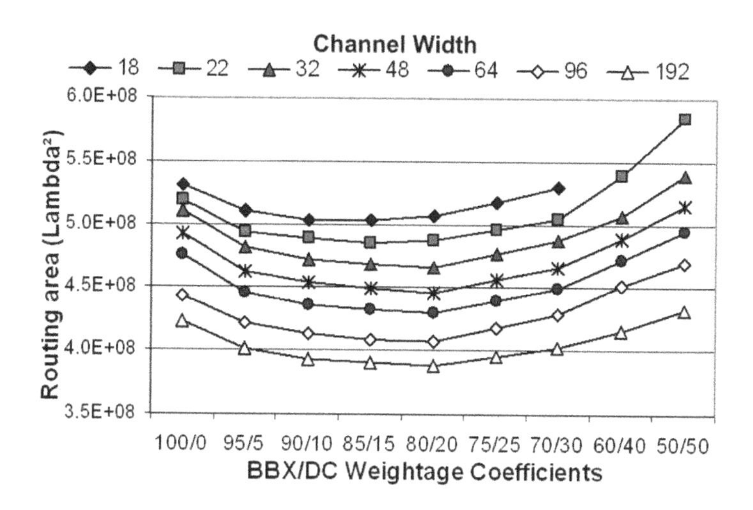

Figure 5.6: ASIF-4 for 17 netlists with varying channel widths, weightage coefficients

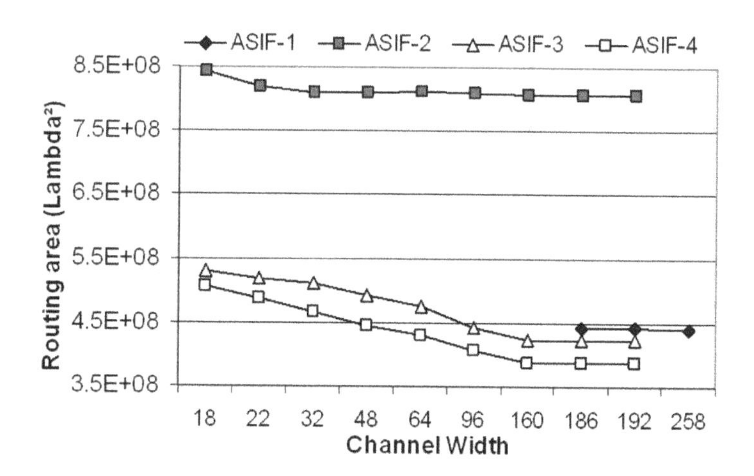

Figure 5.7: Routing area comparison for different ASIFs for 17 netlists

to route 17 MCNC netlists). The other channel widths for ASIF-1 are attained with each netlist routed on slightly increased channel width. For ASIF-2, -3 and -4, the channel width is varied between 18 and 192. It can be seen that routing area difference between ASIF-2 and all other ASIFs is very large. For ASIF-3 and ASIF-4, the routing area decreases as the channel width increases. But on the other hand, Figure 5.8 shows that the number of wires used by ASIF-2 are very low as compared to ASIF-1. For ASIF-3 and ASIF-4, the number of wires increase as the channel width increases. The increase in number of wires is mainly due to preference for the routing cases shown in Figure 5.4(b) and (d). The routing wires can play a pivotal role in the area of ASIF if it dominates the logic area. This can happen if repeated tiles of an ASIF are designed in full-custom, or layout is generated in smaller process technology. In such a case, an ASIF-4 with smaller channel widths can give a compromised solution.

Figure 5.9 shows percentage area distribution of FPGA and ASIF. In an FPGA, only 9.3% area is taken by logic area, whereas the remaining area is taken by the routing area. In ASIFs, the routing area is decreased to such an extent that CLB occupies a very important percentage of the total area; e.g. in ASIF-4 CLB takes 41% of area for a channel width of 192. These results suggest that significant overall gains might be achieved if CLB is also optimized. Section 5.6.3 shows the effect of different LUT sizes on the area of an ASIF. However, just like routing network, the logic resources of an FPGA (i.e. CLBs) can also be optimized by removing unused SRAM resources. If SRAM of a LUT is programmed as '0' by all the netlists, that SRAM can be replaced by a hard-coded '0'. Two similar hard-coded inputs to a two input multiplexor can also be optimized and replaced by a hard-coded bits. This type of optimization will depend upon the total number of netlists used for ASIF generation, and the SRAM programmability of LUTs used by each of the netlists. The SRAM programmability of a netlist can be modified by changing the sequence of input pins of a LUT. For each netlist and each CLB, the best sequence can be determined so that maximum LUT optimization is possible after ASIF generation. However, this kind of optimization is not explored in this work, and is left for future work. The LUTs in an ASIF remain as reprogrammable as in an FPGA. This re-programmability of LUTs in an ASIF can be used to map new application circuits on an ASIF, the technique is discussed later in Chapter 6.

Figure 5.10 compares FPGA and ASIF with changing number of netlists (the order in Table 5.1 is respected). The X-axis presents the number of netlists; where 1 means only "pdc" is used, 2 means "pdc" and "ex5p" are used, and so on. The Y-axis presents the number of times the total area of an ASIF is smaller than the total area of an FPGA. Here ASIFs with maximum channel width are compared with a 68x68 LUT-4 based reference FPGA having channel width of 18. If area occupied by routing wires is not dominant, a LUT-4 based ASIF-4 is 4.39 times or 77.2% smaller than a LUT-4 based 68x68 FPGA for 17 MCNC benchmark circuits. A LUT-4 based ASIF-4 for only 1 MCNC netlist is 10.2 times or 90.2% smaller than 68x68 FPGA.

Figure 5.11 compares routing area of different ASIFs (for maximum channel widths) averaged to ASIF-1. The routing area of ASIF-2 is very large as compared to other ASIFs, and is not compared in the Figure. It should be noted that ASIF-3 is slightly better than ASIF-1. The slight area difference between ASIF-3 and ASIF-1 is due to efficient wire sharing which fa

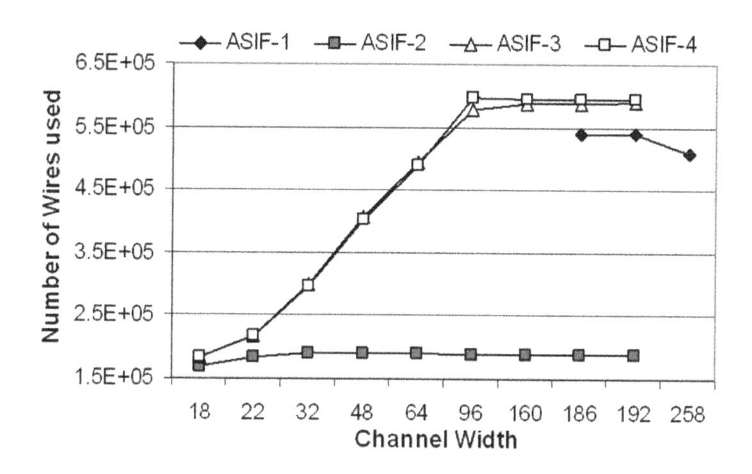

Figure 5.8: Wire count comparison for different ASIFs for 17 netlists

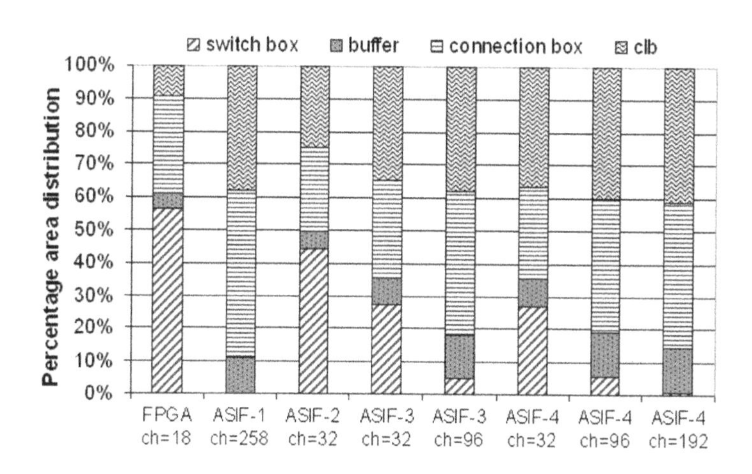

Figure 5.9: Percentage area distribution for FPGAs and ASIFs

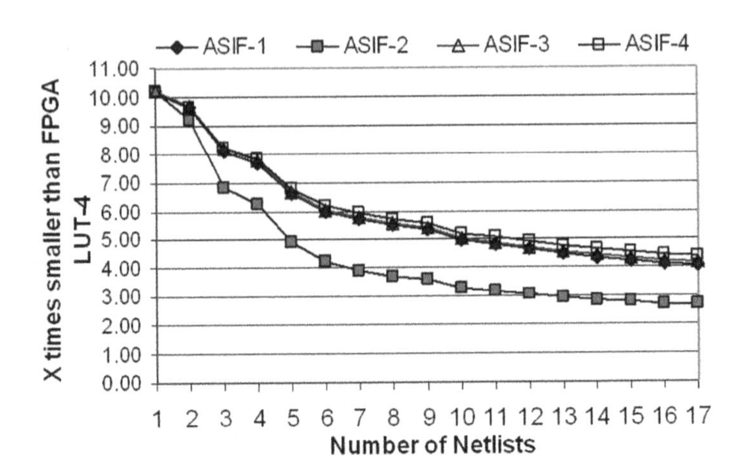

Figure 5.10: FPGA vs. ASIFs with varying number of netlists

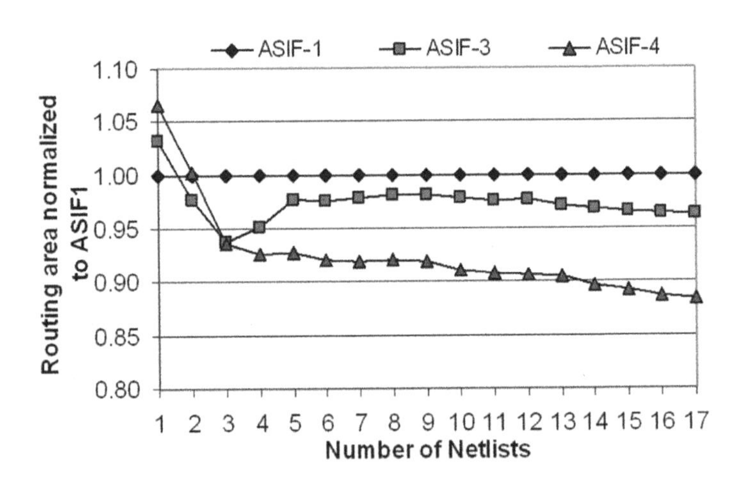

Figure 5.11: Comparison of different ASIFs normalized to ASIF-1

ilitates the use of common switches and wires for several instances of different netlists that appen to be placed on the same blocks of the architecture and drive the same block. Howver, ASIF-1 uses separate switches and wires to connect them. Next, in ASIF-4 both efficient lacement and routing gives a gain up to 12% for 17 netlists. The 4% area gain for ASIF-3 nd 12% area grain for ASIF-4 over ASIF-1 is one of the minor benefits. The major benefit of SIF-4 is the possibility of compromised area/wire-count solution that lies between ASIF-1 nd ASIF-2. For example in Figure 5.7, ASIF-4 area at channel width 32, and in Figure 5.8 SIF-4 wire count at channel width 32 gives a very good compromised solution over ASIF-1 t channel width of 258; i.e. with only 6% increase in the routing area compared to ASIF-1, 2% number of wires have been reduced.

.6.3 Effect of LUT size on ASIF

n order to find the effect of LUT size on ASIFs, experiments are repeated with LUT-2, LUT-3, UT-5 and LUT-6 versions of the MCNC netlists. It is found that the best BBX/DC weighting oefficient ratio used for different LUT sizes is also 80/20. Figure 5.12 compares total area of SIF-4 with several LUT sizes for 17 netlists. Figure 5.13 shows the wire count comparison or ASIF-4 with varying LUT sizes for 17 netlists. It can be seen that a LUT-2 based ASIF equires less area compared to ASIFs with other LUT sizes. However the total number of sed wires increase linearly. Figure 5.14 presents the comparison of FPGA and ASIF-4 with arying number of netlists, varying LUT sizes, and with maximum channel width for ASIF- If the area occupied by wires is not dominant, then a LUT-2 based ASIF-4 is found to e 5.43 times or 81.5% smaller than a LUT-4 based 68x68 FPGA for 17 MCNC benchmark rcuits having channel width of 18. For a single MCNC benchmark circuit, a LUT-2 based SIF-4 is 16 times or (93.75%) smaller than a LUT-4 based 68x68 FPGA having channel width f 18. It is to be noted that 16 times area difference for a single netlist is without logic block ptimization. Figure 5.15 shows the overall area distribution of FPGAs and ASIF-4 with arying LUT sizes. The percentage area occupancy of CLBs in ASIFs has decreased from 2% (in case of LUT-4) to 32% (in case of LUT-2).

or the sake of record, the same experiments are repeated for 20 MCNC benchmark circuits. has been found that a LUT-2 based ASIF-4 for 20 MCNC benchmark circuits is 5.53 times naller than LUT-4 based 92x92 FPGA with channel width of 16. It is to be noted here that hen FPGA size increases from 68x68 to 92x92, the netlist "pdc" is routable with a channel idth of 16 .

.7 Quality Analysis

educed flexibility in ASIFs render many fold area advantages over an FPGA. Previous secons compare different ASIF generation techniques. ASIF-4 generation technique, presented section 5.3.4, gives the least area results. However, these area results are not claimed to be

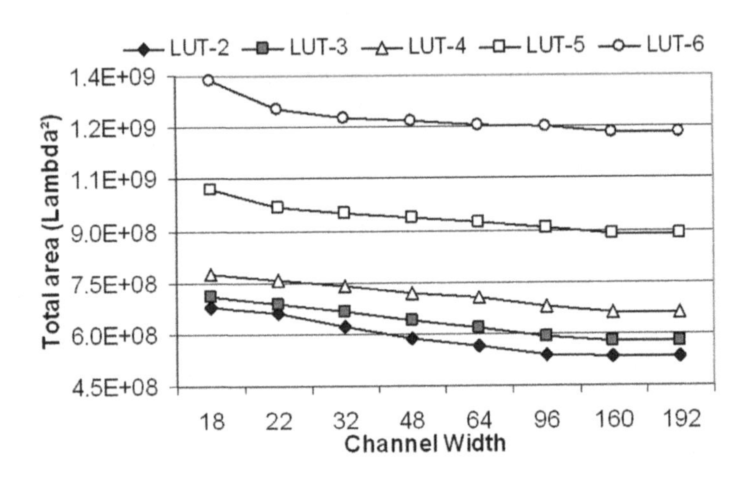

Figure 5.12: Area comparison for ASIF-4 of 17 netlists with varying LUT sizes

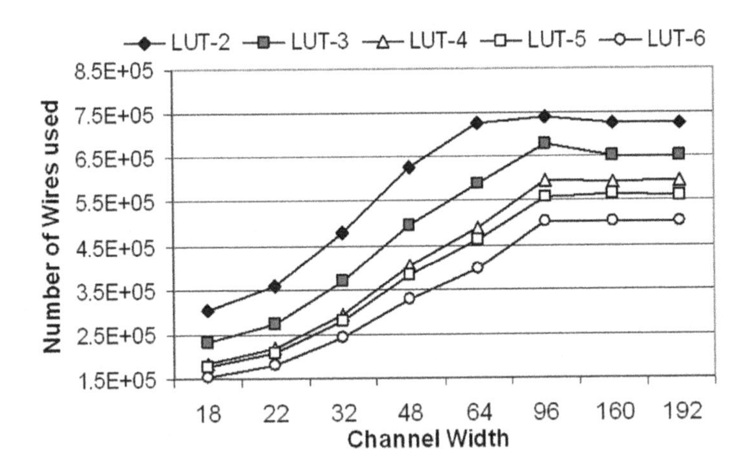

Figure 5.13: Wire count comparison for ASIF-4 of 17 netlists with varying LUT sizes

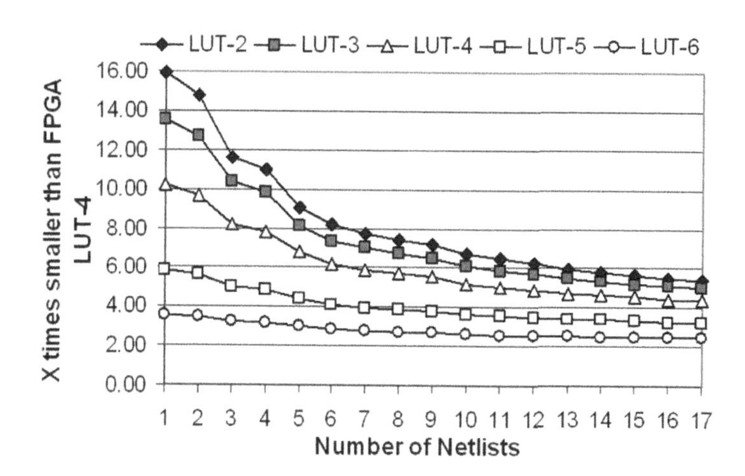

Figure 5.14: FPGA vs. ASIF-4 with varying LUT sizes and number of netlists

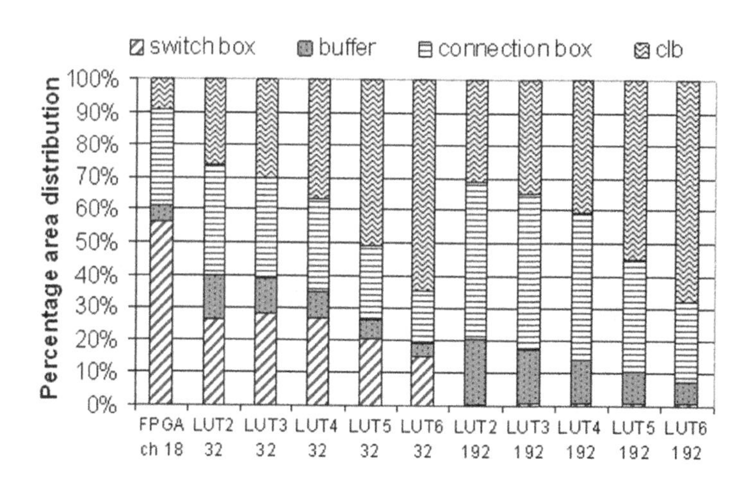

Figure 5.15: Percentage area distribution for FPGAs and ASIF-4 with varying LUT sizes

the best results. An ASIF is generated by using heuristic-based placement and routing algorithms, which do not guarantee to find the best possible solution. The main purpose of this section is to measure the quality of an ASIF. In other words, how good is our ASIF generation methodology?

The quality of an ASIF can be measured by generating an ASIF for a group of netlists for which an ideal ASIF solution is known. The quality of the generated ASIF can then be measured by comparing it with the ideal solution. Thus, if an ASIF is generated for the same netlists, ideally speaking, such an ASIF should not require any switch in the routing channel. This ideal ASIF can be achieved if placement of all the netlists is same, and they use exactly the same routing paths to route their connections.

In this work, two ASIFs are generated to measure the quality of ASIF generation methodology. An ASIF of 10 "tseng" netlists, and an ASIF for 10 "pdc" netlists. "Tseng" and "pdc" represent the minimum and maximum netlist amongst the 17 MCNC benchmarks shown in Table 5.1. The area results for these two ASIFs are shown in Figure 5.16 and 5.17 respectively. The X-axis of the Figures show the number of same netlists used in the generation of ASIF, whereas Y-axis shows the total area of ASIFs. The total area includes the area of logic blocks and routing area. As the number of netlists increase the logic area remains constant, whereas the routing area increases (except for an ideal solution). The ASIF for "tseng" netlists is routed with channel width of 10, whereas the ASIF for "pdc" netlists is routed with channel width of 32. Figure 5.16 and 5.17 compare five different ASIFs as explained below:

- **ASIF generation using Ideal Placement and Routing (Ideal P/R)**

  In this technique, ASIF is generated for the same set of netlists having identical placement and routing. Such an ASIF requires no routing switches. Thus, the area of this ASIF remains constant as the number of netlists increase. This ASIF serves as an ideal solution. Other ASIF generation techniques are compared with this ideal solution. The Ideal placement and routing is achieved manually by using the same placement and routing information for all the netlists.

- **ASIF generation using Normal Placement and Routing (Normal P/R)**

  An ASIF is generated for same set of netlists by using normal placement and routing techniques as used for mapping applications on an FPGA (described earlier as ASIF-2). As all the netlists are similar, it is made sure that the placement solution for these netlists are generated using different random initial placement. Changing the initial placement solution effects the final placement solution. If such a precaution is not taken, similar placement solution will be generated for all the netlists, thus making it an ideal placement solution. It can be seen in Figures 5.16 and 5.17 that the ASIF area using this technique increases linearly as the number of netlists increase. This ASIF serves as a worst case solution.

- **ASIF generation using Efficient Placement and Routing (Eff. P/R)**

  In this technique, netlists are efficiently placed and routed to generate an ASIF (described earlier as ASIF-4). It is found that an ASIF for 10 "tseng" and 10 "pdc" netlists

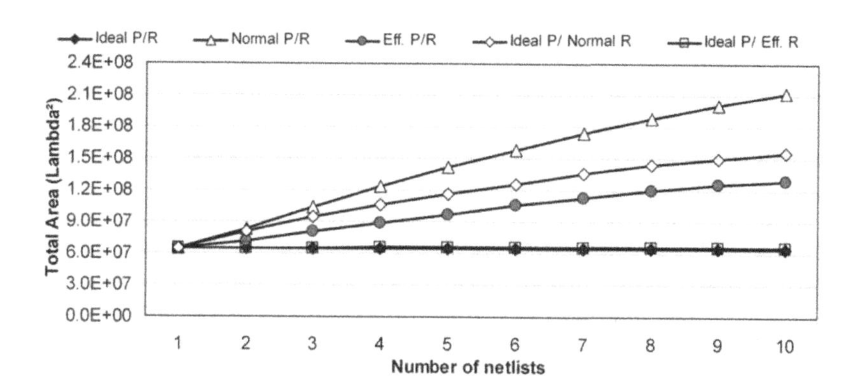

Figure 5.16: ASIF quality measurement for 10 tseng netlists

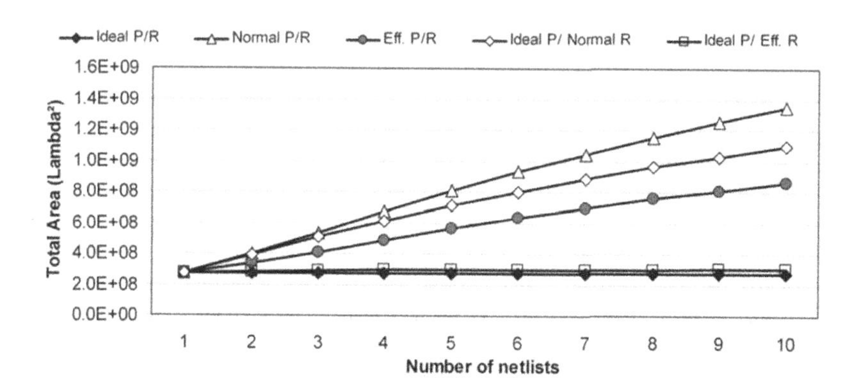

Figure 5.17: ASIF quality measurement for 10 pdc netlists

is almost in the middle of the best and worst solution. The area loss incurred in this ASIF convinced us to further explore the percentage area loss incurred due to placement and routing techniques.

- **ASIF generation using Ideal Placement and Normal Routing (Ideal P/ Normal R)**

  In this case, an ASIF is generated with ideal placement, but with normal routing. As the placement is same for all netlists, normal routing will tend to give same routing solution for all netlists. This is because, the congestion driven routing algorithm does not take any random decisions. The congestion of interconnecting nets in a netlist influence the routing solution. Thus, netlists having same placement will tend to give same routing solution. To counter this effect, the nets of different netlists are randomly shuffled. Thus, the nets of different netlists are routed in different order, which eventually produces different congestion patterns, and thus different routing solutions. We can see in the Figures that an ASIF using ideal placement and normal routing is better than the worst solution, but still much far from the ideal solution.

- **ASIF generation using Ideal Placement and Efficient Routing (Ideal P/ Eff. Routing)**

  In this technique, ASIF is generated using ideal placement and efficient routing. The nets of the netlists are randomly shuffled to add random element in routing algorithm. It can be seen in the Figure that ideal placement with efficient routing produces near ideal results. It affirms that the major loss in ASIF generation is caused by inefficiencies in placement solution.

In conclusion, it has been found that normal placement and routing (ASIF-2) for 10 tseng netlists is 3.3x, and efficient placement and routing (ASIF-4) is 2.0x larger than its ideal ASIF solution. Similarly ASIF-2 for 10 pdc netlists is 4.9x and ASIF-4 is 3.1x larger than its ideal solution. The area of ASIF-2 and ASIF-4 increases as number of netlists increase, whereas area of ideal solution remains constant as the number of netlists increase. Related experiments have revealed that the major area loss in ASIF generation is caused by inefficiencies in placement solution. These results show that there is still a lot of room for improvement in ASIF generation techniques. New placement techniques need to be explored to further improve area of an ASIF.

5.8 Conclusion

This chapter presented Application Specific Inflexible FPGAs (ASIFs) that can implement a set of circuits which will operate at mutually exclusive times. It has been shown that a LUT-2 based ASIF is 5.43 times smaller than a LUT-4 based FPGA for 17 MCNC netlists. The main idea is to design and test a set of application circuits on an FPGA; it can later be reduced to an ASIF for high volume production. A quality analysis of an ASIF generation technique shows that the area of ASIFs can be further optimized by improving efficient placement techniques.

Next chapter describes an ASIF using heterogeneous logic blocks. The limited re-programmability in ASIFs can be exploited to map new or modified applications netlists on ASIFs. A new CAD flow is also presented which can map application circuits on an ASIF.

6

ASIF using Heterogeneous Logic Blocks

\n Application Specific Inflexible FPGA (ASIF) is an FPGA with reduced flexibility that an implement a set of application circuits which will operate at mutually exclusive times. \n ASIF that is reduced from a heterogeneous FPGA is called as a Heterogeneous-ASIF. \ Heterogeneous-ASIF can contain hard-block such as Multipliers, Adders, RAMS or even maller Gates. A set of application circuits are efficiently placed and routed to minimize total outing switches required by the heterogeneous FPGA architecture. Different floor-planning echniques are used to optimize the position of hard-blocks on the FPGA architecture. Later, ll unused routing switches are removed from the FPGA to generate a Heterogeneous-ASIF. 'his work shows that a standard-cell based Heterogeneous-ASIF using Multipliers, Adders nd Look-Up Tables for a set of 10 opencores application circuits is 85% smaller in area than single-driver FPGA using the same type of blocks. This Heterogeneous-ASIF is only 24% rger than the sum of areas of their standard-cell based ASIC versions. If the Look-Up Tables re replaced by a set of repeatedly used hard logic gates (such as AND gate, OR gate, Flip-lops etc), the ASIF becomes 89% smaller than the FPGA and 3% smaller than the sum of SICs. The area gap between ASIF and sum of ASICs can be further reduced if repeatedly sed groups of standard-cell logic gates in an ASIF are designed in full-custom. One of the ıajor advantages of an ASIF is that just like an FPGA, an ASIF can also be reprogrammed to xecute new or modified circuits, but at a very limited scale. A new CAD flow is presented map application circuits on an ASIF.

. Parvez and H. Mehrez, *Application-Specific Mesh-based Heterogeneous FPGA Architectures*,
OI 10.1007/978-1-4419-7928-5_6, © Springer Science+Business Media, LLC 2011

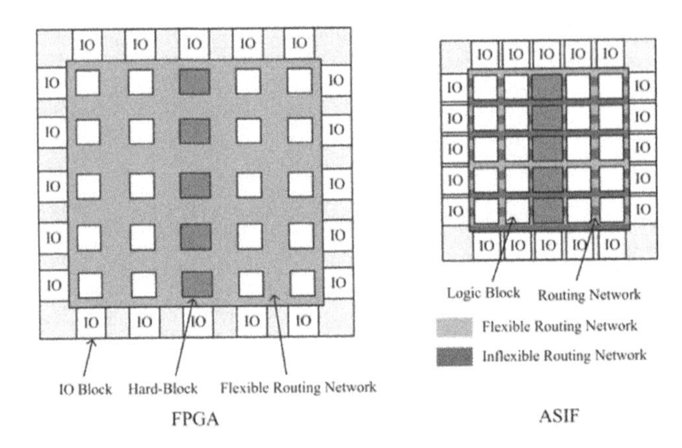

Figure 6.1: An FPGA and an ASIF using Heterogeneous logic blocks

6.1 ASIF Generation Technique

A heterogeneous ASIF is reduced from a heterogeneous FPGA architecture which may contain hard-blocks such as Multipliers, Adders and RAMs etc. Figure 6.1 illustrates an ASIF reduced from an FPGA containing heterogeneous blocks. A group of netlists are placed and routed on the FPGA architecture. Different automatic floor-planning techniques are used to optimize the position of different blocks on the architecture. These floor-planning techniques are already discussed in Chapter 3. After floor-planning of the blocks, netlists are placed and routed on the FPGA architecture. All unused switches are removed from the architecture to generate an ASIF.

A heterogeneous FPGA architecture is initially defined using an architecture description file. BLOCKS of different sizes are defined, and later mapped on a grid of equally sized SLOTS. The PLACER maps multiple netlists together to get a single architecture floor-planning for all netlists. The PLACER moves an instance of a netlist from one BLOCK to another, moves a BLOCK from one SLOT position to another, or rotates a BLOCK around its own axis. After each operation, the placement cost is updated for all disturbed NETS. Depending on cost value and annealing temperature, the operation is accepted or rejected. The area of an ASIF can further decrease if netlists are efficiently placed and routed on the FPGA. Efficient placement and routing techniques are exactly the same as discussed in section 5.3. However, necessary changes are performed to handle heterogeneous blocks. Efficient placement tries to place the instances of different netlists in such a way that minimum routing switches are required in an FPGA. Later, efficient routing encourages different netlists to route their NETS on an FPGA with maximum common routing paths. After all netlists are efficiently placed and routed on FPGA, unused switches are removed from the architecture to generate a heterogeneous ASIF.

6.2 Experimentation and Analysis

6.2.1 Opencores benchmark circuits

This chapter uses a set of 10 opencores circuits to compare different heterogeneous ASIFs. These opencore circuits are synthesized in Synopsys [Synopsys, 2010] using the symbolic standard cell library (SXLIB). Their ASIC layout is performed to get the ASIC size of each of these opencores circuits. The synthesized netlists are passed through the software flow (described in section 3.4) and converted into .NET format. Four different versions of opencores netlists are generated as shown in Table 6.1 & 6.2. The number of multipliers, adders and IOs remain the same in all the four netlists. However these netlists differ by the type of logic blocks. Three versions of the netlists use different sizes of LUTs, as shown in Table 6.1. Fourth version of netlist uses a set of commonly used gates, as shown in Table 6.2. Finding the right mix of gates is not an easy task. This work attempts three different set of gates. Later, the best set of gates are selected for experiments. The three set of gates are as follows: (i) All the netlists are synthesized using the minimum type of gates (which include only nand gates, zero gates, buffers and Flip-Flops), (ii) All the netlists are synthesized using maximum type of gates i.e. any gate required from the standard-cell library, and (iii) All the netlists are synthesized using only commonly used gates required by all the netlists. The set of commonly used gates are manually selected from the synthesis of netlists done using maximum type of gates (as explained in (ii)). The ASIFs generated using the commonly used gates gives minimum area results as compared to the ASIFs generated using minimum and maximum type of gates.

First three versions of the 10 opencores circuits using LUT-2, LUT-3 and LUT-4 (along with multipliers, adders and IOs) are presented in Table 6.1. Table 6.2 shows the 4th version of the same netlists with the commonly used gates. The last row of the Tables gives the maximum FPGA architecture requirement that can map any of the 10 opencores circuits. Although these 10 netlists use different sizes of multipliers and adders, their common maximum size is selected for an FPGA/ASIF. It can also be noticed that due to the nature of the netlists (mainly due to the number of Flip-Flops), there is no major difference between the LUT-4 version and LUT-3 version of the netlist. However, such a case is totally netlist dependent, and cannot be generalized for all the netlists.

6.2.2 Results and Analysis

ASIF generation techniques are applied on four different versions of 10 opencores benchmark circuits shown in Table 6.1 and 6.2. Experiments are performed with different floor-planning techniques, and with various routing resources. ASIF generation techniques presented in section 5.3 are used here. Figure 6.2 shows area comparison of different ASIFs, and Figure 6.3 shows the number of wire segments used by different ASIFs. The X-axis in both of these figures shows ASIFs generated using 4 versions of the netlist, with different floor-planning techniques. The letters ‘C’, ‘M’ and ‘MR’ represent the type of floor-planning

Index	Netlist Name	Mult 16x16	Add 20x20	In	Out	LUT-4	LUT-3	LUT-2
		(Common Elements)				Version-lut4	Version-lut3	Version-lut2
1.	cf_fir_24_16_16	25	48	418	37	2149	2149	2149
2.	rect2polar	0	52	34	40	1328	1497	3657
3.	polar2rect	0	43	18	32	803	803	2280
4.	cfft16x8	0	26	20	40	1511	1937	3282
5.	fm_receiver	1	20	10	12	910	1004	1266
6.	fm_receiver2	1	19	9	12	1308	1508	1987
7.	cf_fir_7_16_16	8	14	146	35	638	638	638
8.	lms	10	11	18	16	940	965	970
9.	rs_encoder	16	16	138	128	537	537	537
10.	cf_fir_3_8_8	4	3	42	18	159	159	159
	Maximum	25	52	418	128	2149	2149	3657

Table 6.1: Netlist Version-lut4,-lut3,-lut2

Netlist Name	Mult 16x16	Add 20x20	In	Out	mux 2	inv 1	an12 2	nor 2	buf 1	and 2	nmux 2	nand 2	zero 1	one 1	Flip -Flop
1. cf_fir_24_16_16	25	48	418	37	0	0	400	0	0	0	0	0	25	10	2116
2. rect2polar	0	52	34	40	962	151	68	202	40	94	0	356	1	18	1113
3. polar2rect	0	43	18	32	736	17	5	2	32	0	1	5	31	14	741
4. cfft16x8	0	26	20	40	165	587	47	883	26	75	256	735	37	10	774
5. fm_receiver	1	20	10	12	0	124	374	363	0	35	10	408	22	17	374
6. fm_receiver2	1	19	9	12	0	205	42	868	234	40	1	566	21	19	248
7. cf_fir_7_16_16	8	14	146	35	0	0	128	0	0	0	0	0	8	11	619
8. lms	10	11	18	16	0	40	65	5	16	13	10	40	22	10	912
9. rs_encoder	16	16	138	128	0	136	0	8	128	136	0	8	33	16	128
10. cf_fir_3_8_8	4	3	42	18	0	0	32	0	0	0	0	0	8	3	148
Maximum	25	52	418	128	962	587	400	883	234	136	256	735	37	19	2116

Table 6.2: Netlist Version-Gates

used (as described in section 3.5). 'C' stands for Column-based floor-planning. 'M' stands for floor-planning achieved through Move operation, where blocks can be jumped or translated. 'MR' stands for floor-planning achieved through Move and Rotate operation. Y-axis in Figure 6.2 shows the total area of ASIFs in Lambda square. Y-axis in Figure 6.3 shows the total number of wire segments (routing tracks) used by different ASIFs. It can be noted in Figure 6.2 that the area of an ASIF decreases as the channel width (routing resource) increases. This is because, the availability of more routing resources increase the probability to prefer case 5.4(b), (c) and (d), which in turn increases the probability to avoid 5.4(a); thus less number of switches are used. However as the channel width increases, the number of wires also increases (as shown in Figure 6.3). The number of wires can play a pivotal role in the area of ASIF if it dominates the logic area. In such a case, an ASIF with smaller channel width can give a trade-off solution.

From Figure 6.2, it can be noticed that ASIF areas come in the following order : ASIFArea (LUT-4) > ASIFArea (LUT-2) > ASIFArea (LUT-3) > ASIFArea (Gates). This effect is mainly due to the area of logic blocks in ASIFs. Because of the nature of netlists (shown in Table 6.1), the maximum number of LUT-4 in version-lut4 are same as that of LUT-3 in version-lut3. Thus ASIF using version-lut3 is smaller than ASIF using version-lut4. On the other hand, the number of LUT-2 in version-lut2 are much greater than the number of LUT-3 in version-lut3. Thus, despite LUT-2 being smaller in size than LUT-3, the total logic area in an ASIF using version-lut2 becomes more than the logic area occupied by the ASIF using version-lut3. The total logic area of ASIF in version-gates is much smaller than logic area in version-lut3; however a gate based ASIF uses more routing resources than used by LUT-3 based ASIF. But, eventually gate based ASIF gives the best result in terms of area.

The effect of floor-planning can also be noticed in Figure 6.2. ASIFs for LUT-4, LUT-3 and LUT-2 based netlists show only a slight area gain (i.e. between 0.5% and 6%) with Move ('M') and Move/Rotate ('MR') floor-planning over Column floor-planning ('C'). However, in case of gates version, a considerable gain (i.e. between 8% to 21%) is achieved with Move floor-planning 'M' over Column floor-planning ('C'). This is because the gates version of netlists uses 13 different types of blocks, whereas LUT based versions use only 3 types of blocks. The floor-planning of 13 types of blocks in columns greatly restricts the placements of different instances of a netlist, which in turn requires more routing resources to connect them with each another. The Move/Rotate floor-planning did not show much better area results over Move floor-planning.

Figure 6.4 and 6.5 compare three different ASIF generation techniques. The comparison is done for 10 netlists with FPGA floor-planning attained through block Move/Rotate operation. These techniques differ in the placement and routing of netlists on FPGA, before being reduced to ASIF. The section (i) in Figure 6.4 and Figure 6.5 uses normal placement and routing (i.e. ASIF-2 technique as presented in section 5.3.2), (iii) uses efficient placement and routing (i.e. ASIF-4 techniques as presented in section 5.3.4). Section (ii) uses normal placement, whereas routing of each netlist is performed using different channel tracks (i.e. ASIF-1 technique as presented in section 5.3.1). ASIF-1 technique does not share switches in the switch boxes and the routing tracks amongst different netlists, only logic blocks and connec-

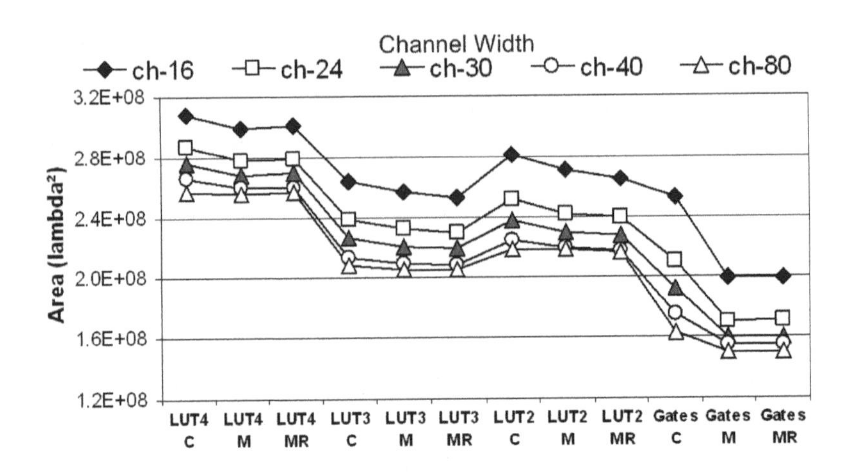

Figure 6.2: Area comparison of ASIFs with different netlists/floor-plannings

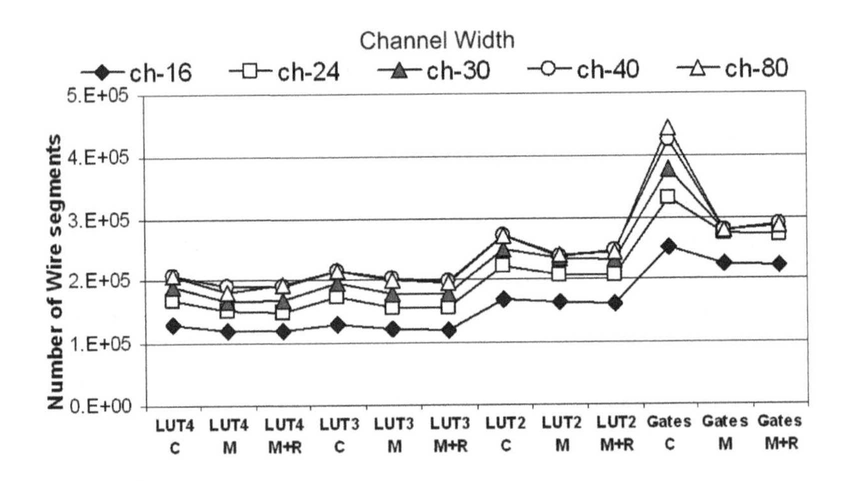

Figure 6.3: Wire count comparison for different ASIFs

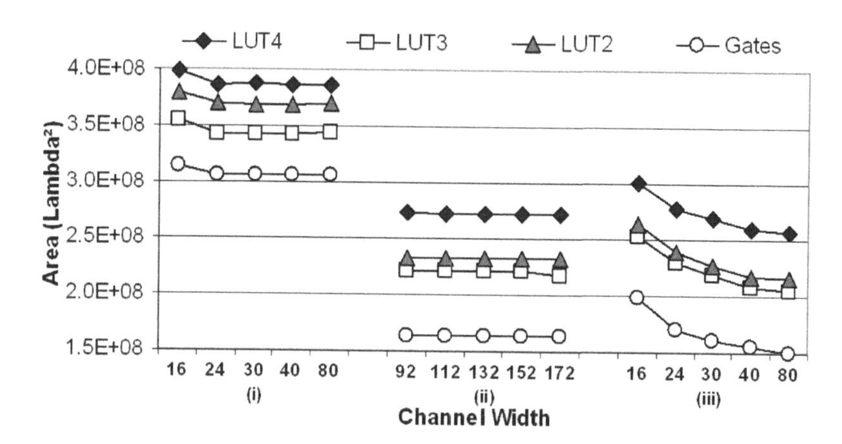

(i) Normal Placement/Routing (ii) No wire sharing
(iii) Efficient Placement/Routing

Figure 6.4: Area comparison for ASIFs with different Placement/Routing

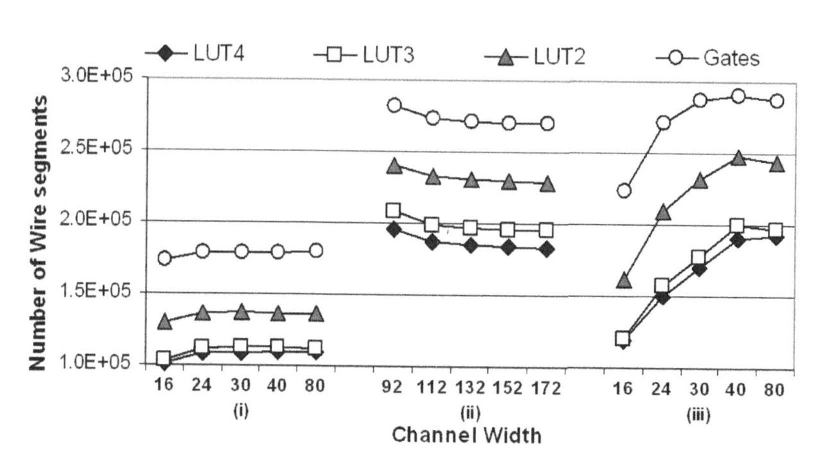

(i) Normal Placement/Routing (ii) No wire sharing
(iii) Efficient Placement/Routing

Figure 6.5: Wire count comparison for ASIF with different Placement/Routing

tion boxes are shared. Consequently, the routing channel width used by ASIF-1 is the sum of the channel widths used by all the netlists. On the other hand ASIF-2 and ASIF-4 share routing channels for all netlists.

The area of an ASIF shown in Figure 6.4 (iii) is 33 to 51 percent better than Figure 6.4 (i), and 6 to 9 percent better than Figure 6.4 (ii) for four different netlist versions at maximum channel widths. The slight improvement of (iii) over (ii) is due to efficient placement, which places several instances of different netlists on the same blocks of the architecture; later efficient routing facilitates the use of common switches and wires to drive the same blocks; thus reducing the number of switches.

Figure 6.5 shows that the number of wires used in (i) are very less as compared to (ii) and (iii). The wires used in (ii) remains relatively constant, and the wires used in (iii) change with different channel widths. In an ASIF for large number of netlists, the wires used by (ii) can become so huge that the overall area of chip becomes wire dominant. In such a case, an ASIF shown in (iii) can give a trade-off solution using smaller channel widths.

Figure 6.6 compares the sum of areas of ASIC with ASIFs for varying number of netlists (the order of netlists in Table 6.1 is respected). The best achieved ASIFs (generated with maximum channel width and Block Move/Rotate floor-planning) are compared here with the sum of areas of ASICs. The X-axis represents the number of netlists used in the experiment; 1 means that only "cf_fir_24_16_16" is used, 2 means that "cf_fir_24_16_16" and "rect2polar" are used, and so on. The Y-axis represents the area in symbolic units (Lambda square). If area occupied by routing wires is not dominant, a LUT-3 based ASIF for 10 netlists is only 24% larger than the sum of areas of ASICs, whereas a gate based ASIF is 3% smaller than the sum of areas of ASICs for 10 netlists. The sum of areas of ASICs is slightly smaller than the gate based ASIF for 7 or less netlists. Whereas for more than 7 netlist, the gates based ASIF becomes smaller than the sum of areas of ASIC. Comparison with an FPGA (not shown in Figure) shows that a LUT-3 based ASIF is 85% and a gate based ASIF is 89% smaller than a Heterogeneous LUT-3 based, single-driver, unidirectional FPGA using a channel width of 12.

Figure 6.7 shows the percentage area distribution in FPGA and ASIFs. In a LUT-3 based FPGA, using a channel width of 12, only 9.3% of the area is taken by logic area (CLBs, Multipliers and Adders), whereas the remaining area is taken by the routing area (switch boxes, connection boxes and buffers). In ASIFs, the routing area is decreased down to such an extent that the logic area occupies a very important percentage of the total area; in gates version of ASIF, logic area takes 50% of area for a channel width of 80; in LUT-3 version of ASIF, logic area takes 64% of area for a channel width of 80.

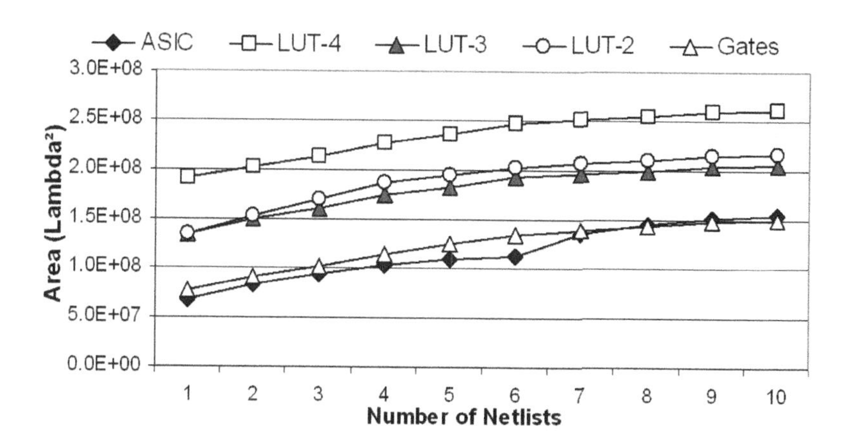

Figure 6.6: ASIC vs. ASIFs with varying LUT sizes and number of netlists

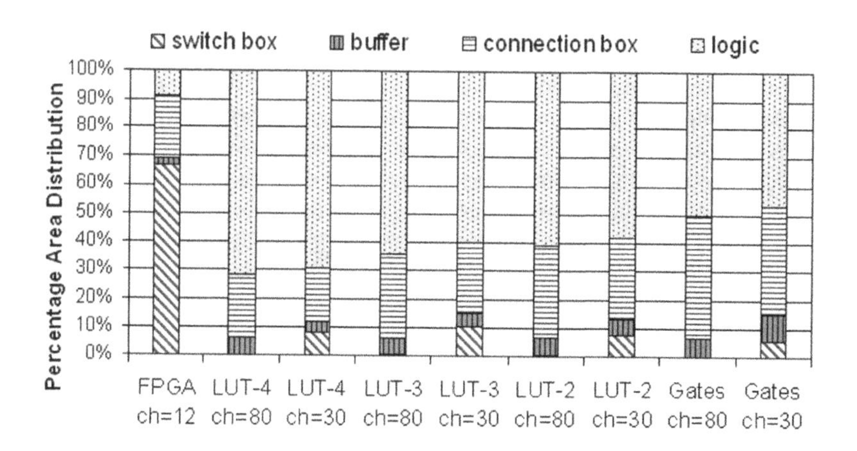

Figure 6.7: Percentage area distribution for FPGAs and ASIFs

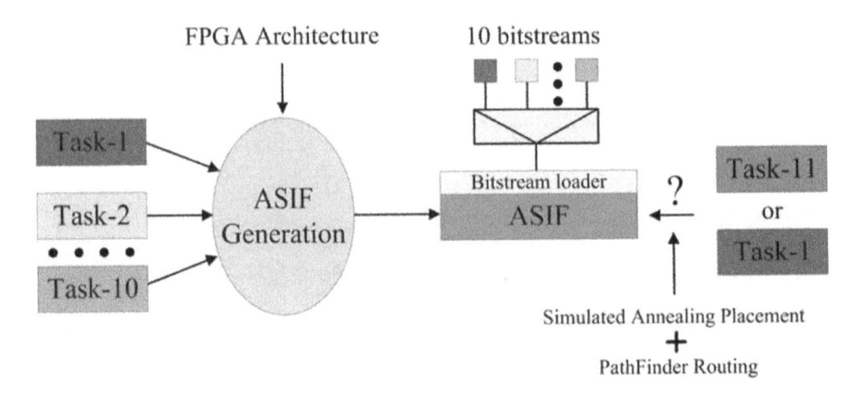

Figure 6.8: Can an application be mapped on an ASIF?

6.3 Mapping application circuits on ASIF

6.3.1 Introduction

An ASIF retains limited routing resources of an FPGA to execute only a known set of circuits. However, these limited resources in an ASIF might be exploited to execute any new circuit. Though an ASIF is unable to map the same variety of circuits as an FPGA does, it might be used to map some elementary circuits, or slightly modified versions of the same circuits used in the generation of an ASIF. So, is it really possible to map new or even the same application circuit for which an ASIF is generated? Figure 6.8 illustrates this phenomenon. One of the major problems to map a new netlist on an ASIF is to find an appropriate placement with sufficient routing paths to route all signals. It is not sure if such a placement solution exists in an ASIF or not. Even if few routable placement solutions exist, the currently used heuristic based placement algorithms do not guarantee to find a placement solution from a solution space that contains few or maybe only one placement solution. In this regard, we attempted the simulated annealing based placement algorithm to re-place the same set of netlists on an ASIF which are used in its generation. The aim of the simulated-annealing placement algorithm was to find a placement of the new netlist without increasing the "Driver Count" cost (see section 5.3.4) of the ASIF. However, the algorithm was unable to find a routable placement solution in limited time.

6.3.2 Algorithm

This section presents an elementary version of a branch-and-bound [Lawler and Wood, 1966] based placement algorithm which is able to re-place the same set of circuits on their own ASIF. The same algorithm is later used to map a simple new netlist on an ASIF. The main idea is to attempt placement of all instances of a netlist on all possible BLOCK positions of the ASIF. Although, limited routing resources decrease inter-BLOCK connectivity in an

ASIF; attempting all possible combinations can however be an extremely time consuming task. For that reason, the placement combinations for instances of a netlist are systematically decreased. The algorithm also ensures that the placement of a driver and its receiver instances are only attempted on interconnected BLOCKS.

The algorithm initially extracts BLOCK connectivity from an ASIF (i.e. which driver PINS of a BLOCK are routable to which receiver PINS of a BLOCK). Connectivity information is also extracted for the netlist that is attempted to be mapped on the ASIF. The connectivity information of the ASIF and the netlist are used to generate an exhaustive list of BLOCKS (called here as maplist) on which each instance of a netlist can be mapped. For each instance 'i', the maplist is further reduced by recursively attempting to place its driver and receiver instances on the BLOCKS found in their own maplist and validating if routing connection exists between the placed instance 'i' and its placed driver and receiver instances. A BLOCK is removed from the maplist of an instance 'i' if any of its Nth level driver and receiver is unable to find any routable BLOCK position. Once the maplist is finalized for all the netlist instances, a recursive, tree-pruning based algorithm attempts placement of all instances in a netlist over all their possible BLOCK positions on an ASIF.

The proposed algorithm is shown in Figure 6.9. The *BlockArray* contains the BLOCK connectivity information of an ASIF. The *InstArray* contains the instance connectivity information of the target netlist that is to be mapped on the ASIF. The *BlockMapList* contains list of all BLOCKS on which a particular instance can be mapped. The function *ComputeInitialInstBlockMapList* (at line 00 of the algorithm in Figure 6.9) computes an initial list of BLOCKs on which each instance of the target netlist can be mapped. A BLOCK is added in the *BlockMapList* of an instance if (i) the type of the instance 'i' and the BLOCK 'b' are identical, (ii) the type of the driver and receiver instances of the instance 'i' is the same as the type of the driver and receiver BLOCKS of the BLOCK 'b', and (iii) the number of driver and receiver PINS of the instance 'i' are equal or less than those of the BLOCK 'b'. The *BlockMapList* is further reduced by the function *ReduceInstBlockMapList* (line 01), which is explained later in this section.

Once the *BlockMapList* is finalized for all instances, the netlist is arranged in ascending order according to the size of *BlockMapList* (line 02) and is then placed using the function *PlaceAllInstances* (line 05). The *PlaceInstance* function is called recursively for all instances in the netlists. For every instance, each block position is tested (line 16). A BLOCK position is skipped if a block is already occupied (line 18) or the position is not possible (line 20) (i.e. line 30) the current BLOCK on which the instance is tested has no available connection with any of its driver or receiver instances already placed on some other BLOCK). If a placement for an instance is found, the placement for next instance is called recursively (line 24). Once the placement is found for all instances, a placement file is generated (line 11), and routing is performed (line 12). The generated placement only guarantees that routing connections exist between all the interlinked instances of the netlist. However, due to conflict of routing resources, the routing might fail. So the process continues until a valid routing solution is found. The *ReduceInstBlockMapList* uses a function similar to *PlaceInstance*. However unlike the *PlaceInstance* function (which is recursively called for instances in a sequence until all

Let: ***BlockArray[m]*** *is the list of all the blocks in an ASIF,* ***InstArray[n]*** *is the list of instances of the target netlist that is to be mapped on ASIF,* ***InstArray[n]–>BlockMapList*** *is the list of blocks on which the instance 'n' can be mapped.*

```
00  ComputeInitialInstBlockMapList();
01  ReduceInstBlockMapList();
02  OrderInstInAscendingOrderOfMapListSize();
03  PlaceAllInstances();
04
05  void PlaceAllInstances() {
06    placeInstance(0);
07  }
08
09  void PlaceInstance(InstNo) {
10    If(InstNo == totalInstances) {
11      Generate_placement_file()
12      Try_Pathfinder_routing_algorithm()
13      return;
14    }
15
16    For i = 0 to InstArray[InstNo].BlockMapList.size() {
17      Block = InstArray[InstNo].BlockMapList[i]
18      if(Block–>Busy() == true)
19        continue;
20      if(isPositionPossible(Block, InstNo)==false)
21        continue;
22      InstArray[InstNo].block_position_found = Block
23      Block–>SetBusy(true);
24      PlaceInstance ( InstNo + 1)
25      Block–>SetBusy(false);
26      InstArray[InstNo].block_position_found = NULL;
27    }
28    return
29  }
```

(continued...)

```
30 bool isPositionPossible(Block, InstNo) {
31   For i = 0 to All the pin classes of InstNo {
32     pinclass = i
33     For r = 0 to All the receiver instances of InstNo {
34       rcv_ins = receiver Instance
35       rcv_class = receiver pin class
36       If the rcv_ins is already mapped on a block {
37         rcvblock = rcv_ins–>block_position_found
38         // If Block–>pinclass can reach rcvblock–>rcv_class
39         flag = ifAcanReachB(Block,pinclass,rcvblock,rcv_class)
40         if(flag == false)
41           return false;
42       }
43     }
44
45     For d = 0 to All the driver instances of InstNo {
46       drv_ins = driver Instance
47       drv_class = driver pin class
48       If the drv_ins is already mapped on a block {
49         drvblock = drv_ins–>block_position_found
50         // If drvblock–>drv_class can reach Block–>pinclass
51         flag = ifAcanReachB(drvblock,drv_class,Block,pinclass)
52         if(flag == false )
53           return false;
54       }
55     }
56   }
57 }
```

Figure 6.9: Algorithm to place netlist on ASIF

Index	Netlist Name	ASIF (LUT-3)	ASIF (Gates)	FPGA (LUT-3)
1.	cf_fir_24_16_16	180	151	445
2.	rect2polar	11589	173	263
3.	polar2rect	182	117	105
4.	cfft16x8	∞	122	340
5.	fm_receiver	922	81	103
6.	fm_receiver2	1083	51	183
7.	cf_fir_7_16_16	66	60	57
8.	lms	∞	357	69
9.	rs_encoder	193	69	58
10.	cf_fir_3_8_8	39	34	11
11.	scan_path(900 CLBs)	202	-	-

Table 6.3: Placement Execution Time (in seconds)

instances are placed); this function is individually called for each instances, and recursivel called for its driver and receiver instances.

6.3.3 Results and Analysis

The proposed algorithm is used to find a routable placement solution for 10 opencores ci cuits which are used to generate the ASIF. Execution time for mapping a LUT-3 and gate version netlists on their respective ASIFs is shown in Table 6.3. For the sake of compar son, the execution time of simulated annealing based placement algorithm to map netlis on a LUT-3 based FPGA is also shown in the table. The execution time of this algorith mainly depends on the number of possible BLOCK positions on which each instance can b mapped. A generalized BLOCK such as a Look-Up Table tends to give more BLOCK pos tions to a netlist instance than a specialized BLOCK such as Multiplier or AND gate etc. Fo this reason, a LUT-3 based ASIF takes more execution time compared to a gate based ASI Due to the same reason, the algorithm is unable to find a routable solution for two LUT- version netlist in limited time. However, an ASIF with a generalized BLOCK is more suitab to map new circuits. A simple scan-path circuit passing through 42% of the CLBs has als been successfully mapped on an LUT-3 version ASIF for 10 opencores circuits. A scan-pat circuit passing through all CLBs is not possible due to limited inter-CLB connections. Ma ping of more complex circuits is not considered in this work, it is one of the major futu direction for ASIF exploration. Target netlist modifications can be performed to map ne netlists on an ASIF. The netlist modifications may include addition of buffers and instan replication to better adapt the new netlist according to ASIF characteristics. Besides, the still remains a lot of room for optimization at the algorithmic and implementation lev These optimizations can further decrease the execution time to map new or modified circu on an ASIF.

6.4 Conclusion and Future Work

This chapter presented Application Specific Inflexible FPGAs (ASIFs) that can implement a set of circuits which will operate at mutually exclusive times. It has been shown that a Heterogeneous ASIF containing multipliers, adders and LUT-3 is 85% smaller than a LUT-3 based FPGA, and only 24% larger than the sum of ASIC for 10 opencores netlists. The Heterogeneous ASIF containing multipliers, adders and gates is 89% smaller than a LUT-3 based FPGA, and 3% smaller than the sum of ASIC for the 10 opencores netlists. One of the main advantages of ASIFs over ASICs is that it can be reprogrammed (although at a limited scale); this feature can be used to perform minor modifications of netlists after manufacturing, or maybe used to map new netlists. This work also presented a basic algorithm for mapping circuits on an ASIF. The execution time of this algorithm can be further reduced by improving the algorithm and its implementation. More complex circuits can also be mapped on an ASIF by modifying the netlists (through buffer insertion, instance replication etc) to deal with ASIF resource limitations.

Next chapter describes a hardware generation method for different ASIF architectures supported by the ASIF exploration environment. The same 10 opencores benchmark circuits are used to generate VHDL model of different ASIFs. Their layout is performed and compared with the layout area of sum of ASICs.

7

ASIF Hardware Generation

his chapter presents an automated method of generating hardware description of ASIF rchitectures. The ASIF hardware generator is integrated with the ASIF exploration environ- nent. By doing so, all ASIF architectural parameters that are supported by the exploration nvironment are automatically supported by the VHDL generator. A bitstream generator s also integrated with the exploration environment. It generates programming bitstream or each netlist mapped on ASIF. These bitstreams can be individually programmed on the SIF to execute different application circuits exclusively. The VHDL model of an ASIF is imulated using Synopsys. Bitstreams of different application circuits are programmed and ested on ASIF. The VHDL model is later passed to Cadence Encounter to generate layout of SIFs for 130nm 6-metal layer CMOS process of ST Micro-Electronics.

SIF generation techniques presented in previous chapters show that area of an ASIF de- reases as channel width increases. However, the number of total routing wires increase. he increase in the number of routing wires may make an ASIF core to be wire dominant. hus a lot of empty space (fillers) might be required in an ASIF to make it routable. ASIF youts are generated for 10 opencore circuits. It is found that an increase in channel width icreases routing congestion of ASIF layout, however it remains logic dominant for 10 open- ore circuits.

.1 Introduction and Previous Work

n Application Specific Inflexible FPGA (ASIF) can be generated to support a particular omain or set of application circuits. An ASIF shares logic and routing resources among

. Parvez and H. Mehrez, *Application-Specific Mesh-based Heterogeneous FPGA Architectures*,
OI 10.1007/978-1-4419-7928-5_7, © Springer Science+Business Media, LLC 2011

different set of circuits. It uses limited amount of flexibility to execute different application circuits at exclusive times. An FPGA-based digital product, which provides multiple functionalities at exclusive times, can be quickly designed and tested on an FPGA. Later it can be reduced to an ASIF to achieve area, performance and volume production gains. However, these specialized ASIFs would become totally impractical if they are designed manually for each group of applications. An automatic hardware generator for ASIFs is very essential to decrease the design cost and time-to-market of application specific reconfigurable devices.

An ASIC can also be designed for a group of application circuits required to be executed at exclusive times. But such an attempt can be very time consuming, as hardware resources need to be manually shared among different circuits. Thus overall design cost and time-to-market of the ASIC further increases. Synthesis tools have some ability to find resource sharing opportunities, but unfortunately they require the designs to be merged into a single design. [Compton and Hauck, 2007] have verified that synthesis tools provided by Xilinx and Synopsys are unable to find sharing opportunities if circuits are simply instantiated in a larger framework. Flattening the circuits might help. However, the synthesis time increases dramatically, and many sharing opportunities are still overlooked by the synthesis tools.

This chapter considers a different layout generation technique than the one presented in Chapter 4. The automatic FPGA layout generator presented earlier is a tile-based layout generator. It generates different tiles of an FPGA that are later abutted together to form the complete layout of FPGA. The tile-based FPGA layout generator has its own advantages and disadvantages. The major advantages is that a detailed architectural netlist and layout is automatically generated. Provisions for manual interventions allow to tailor the layout according to the desired requirements. Moreover, the layout is generated using a symbolic standard-cell library, which allows to migrate the layout to any fabrication process. However, one of the major disadvantages of the previously proposed tile-based FPGA layout generator is that it is not linked with the architecture exploration environment. Any changes done in the architecture exploration environment are not immediately reflected in the layout generator. Moreover, considerable amount of area is wasted when symbolic layout is converted to real layout. This is due to the generic design rules of symbolic standard cell library which allows to migrate symbolic layout to any fabrication process. These disadvantages of the previously proposed FPGA layout generator convinced us to attempt a different layout strategy for ASIFs.

The VHDL generator for an ASIF is integrated with the ASIF exploration environment. The position of different logic blocks (ASIF floor-planning), and the routing graph of the ASIF are used to generate the VHDL model of ASIFs. The major benefit of this approach is that all architectural changes in the exploration environment are directly reflected in the VHDL model. Different size of ASIFs, varying channel widths, different variety of heterogeneous blocks, modification in connection and switch boxes etc. are automatically supported by the VHDL model. If the routing graph represents a complete FPGA rather than an ASIF this VHDL generator can also be used to generate an FPGA. The generated VHDL model is then passed to commercial layout tools, such as Cadence Encounter, for hardware layout generation.

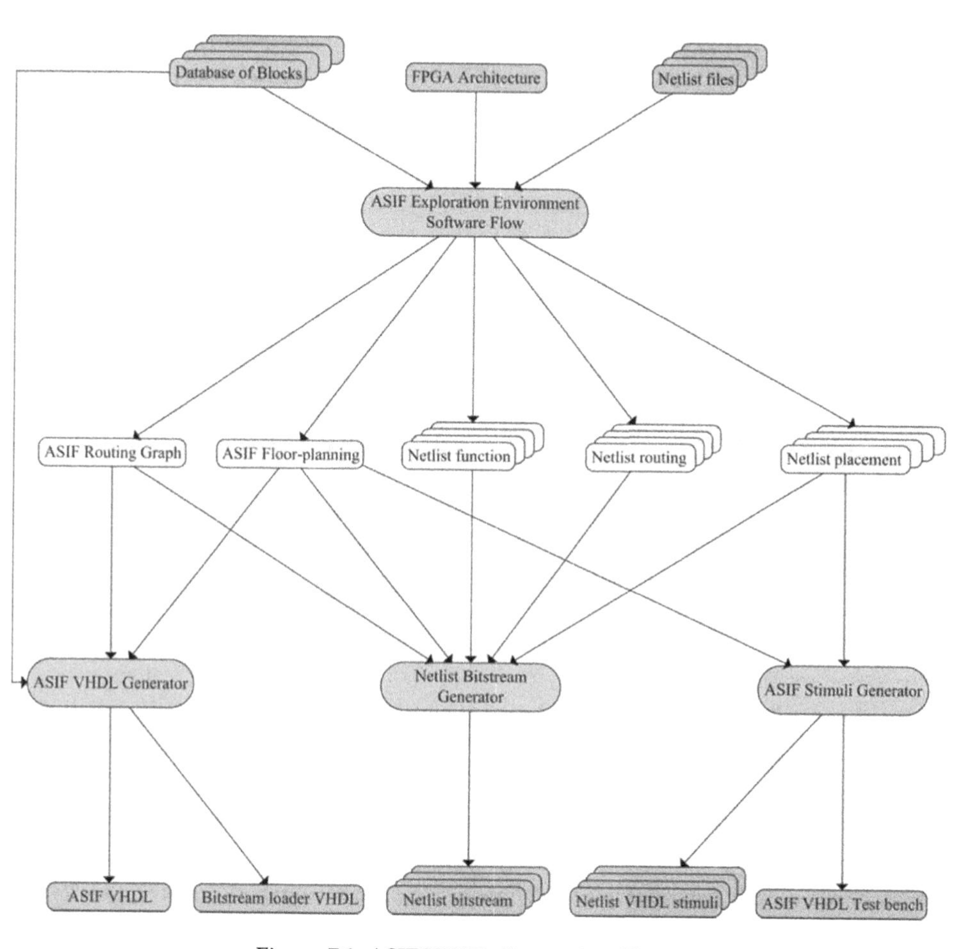

Figure 7.1: ASIF VHDL Generation Flow

7.2 ASIF Generation

The main sections of ASIF generation are described as follows.

7.2.1 ASIF Generation Flow

Figure 7.1 shows the complete ASIF generation flow. The ASIF exploration environment receives three input parameters, (i) A set of application netlists for which an ASIF is required, (ii) architecture description parameters of an FPGA that is later reduced to an ASIF, and (iii) a database of different heterogeneous blocks used in the FPGA architecture. A maximum FPGA architecture is initially defined that can map the given set of application netlists at mutually exclusive times. A software flow efficiently places and routes the netlists on the FPGA architecture. The floor-planning of different heterogeneous blocks on the FPGA architecture is also optimized. After the netlists are mapped on FPGA, all unused routing resources of FPGA are removed to generate an ASIF. The software flow generates placement and routing files for each netlist. A netlist function file is also generated which contains configuration of Look-Up Table(LUT) used by configurable logic blocks (CLBs) of the netlists.

An ASIF is represented by the floor-planning of different blocks and the routing graph connecting these blocks. The routing graph of an ASIF is the union of all the routing paths used by any of the mapped netlists. The ASIF VHDL generator uses the database of different heterogeneous blocks, ASIF floor-planning and ASIF routing graph to generate the ASIF VHDL model. It also generates the VHDL model of a bitstream loader that is required to configure a netlist bitstream on the ASIF. A bitstream generator generates the bitstream for each netlist. The bitstream generator uses the ASIF floor-planning and routing graph, the placement, routing and function information of each netlist to generate netlist bitstream configuration. The execution of different netlists can be switched on the ASIF by loading their respective bitstream. Test bench and stimuli files are also generated which are used to simulate an ASIF for different netlists.

7.2.2 Example Architecture

This chapter mainly describes hardware model (VHDL) generation of an ASIF; the layout is later done through ENCOUNTER. However, the ASIF hardware generator considers an approximate layout scheme to generate an efficient VHDL model. The ASIF VHDL generation, along with an approximate layout scheme, is explained here with the help of an example architecture.

Suppose, an ASIF is required for a set of netlists that can be mapped on a 3x3 homogeneous FPGA architecture. A 3x3 FPGA architecture using unidirectional routing network is shown in Figure 7.2. The layout of a 3x3 FPGA is divided into 16 (i.e 4x4) small tiles. Each tile is represented by (x,y) coordinates, shown on the bottom-left of each tile. A tile consists of a switch box, connection boxes, a CLB, and routing wires on top and right side of the CLB. The

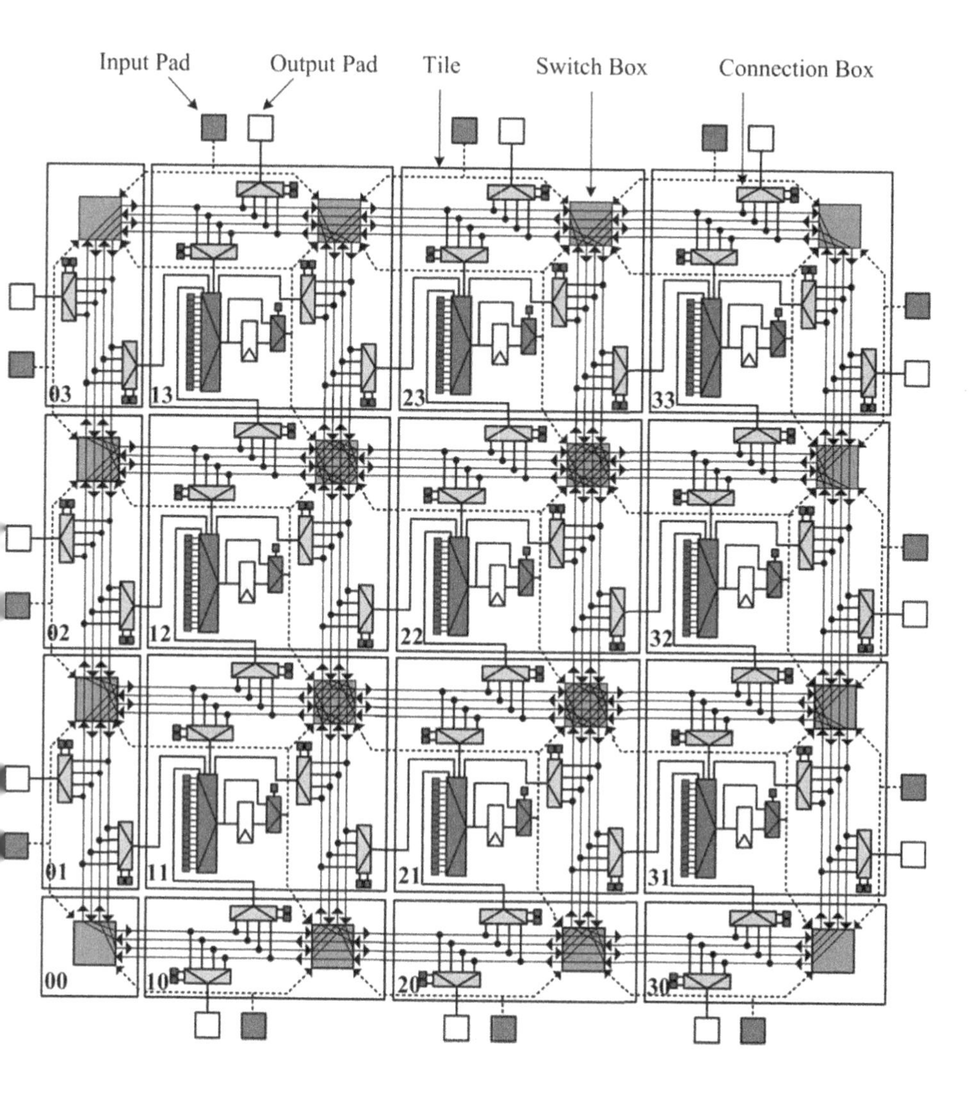

Figure 7.2: A 3x3 FPGA with single-driver, unidirectional routing network

tiles on the left and bottom periphery of FPGA does not contain CLBs. They only contain switch box and connection box. In case of heterogeneous FPGA, a hard-block can occupy multiple tiles.

7.2.3 ASIF Configuration Mechanism

One of the major layout consideration in an ASIF (or even an FPGA) is the programmability of SRAMs. An SRAM is programmed to 0 or 1. A data signal writes a bit to an SRAM if the strobe signal is high. If the strobe signal is low, the data signal is ineffective. The output of an SRAM is used to render reconfigurability to logic or routing resources of an ASIF. The SRAMs are generally placed near different components (generally multiplexors) that require them. This is done to decrease routing congestion in the layout. Eventually, SRAMs are scattered through out the ASIF. It is not at all a viable option to provide distinct data and strobe signals to program these scattered SRAMs. Thus, data and strobe signals are shared between different SRAMs. A homogeneous FPGA architecture generally contains repeatable tiles, with each tile containing same number of SRAMs. However, an ASIF is generated by removing unused sections of switch box and connection box, thus number of SRAMs vary in each tile of an ASIF. The ASIF hardware generator shares the data and strobe signals between varying number of SRAMs in different ASIF tiles.

The bitstream configuration mechanism of a 3x3 ASIF is shown in Figure 7.3. It can be seen that different ASIF tiles can have varying number of SRAMs. These SRAMs in a tile are arranged in words, each word contains maximum eight SRAMs. The data signals are shared by different tiles placed in the same column, and by SRAM words of a tile. The strobe signals are shared by tiles in a row. The number of data signals entering a tile from top, and the number of strobe signals entering a tile from left are selected to be maximum required for any tile in the ASIF. The data signals traverse all the tiles in the entire column, and the strobe signals traverse all the tiles in the entire row of an ASIF. Buffers are added before the data and strobe signals enter the next tile. The number of data and strobe signals entering a particular tile are variable parameters. These parameters can be changed according to the maximum number of SRAMs required in a tile. In Figure 7.3, the maximum number of strobe and data signals entering a tile are 10 and 8 respectively.

A bitstream loader is attached with an ASIF. The loader comprise of top and left decoders, and a set of scan registers. The scan registers receive the column number, row number, word number and data bits through "Scan In" and "Scan out" pads. The main purpose of scan path is to reduce the IO PAD requirement for programming an ASIF chip. The top column decoder receives a column number to turn on one column signal at a time. The left decoder receives the row number and word number to turn on a strobe signal when write enable signal is turned on. An AND gate is placed before each SRAM word. If the horizontal strobe signal and vertical column signal are high, the strobe is set as high for the selected SRAM word. When strobe is high, vertical data signals are written on the SRAM word. In Figure 7.3 column number 2, row number 3 and word number 0 are used to write eight data bits on the

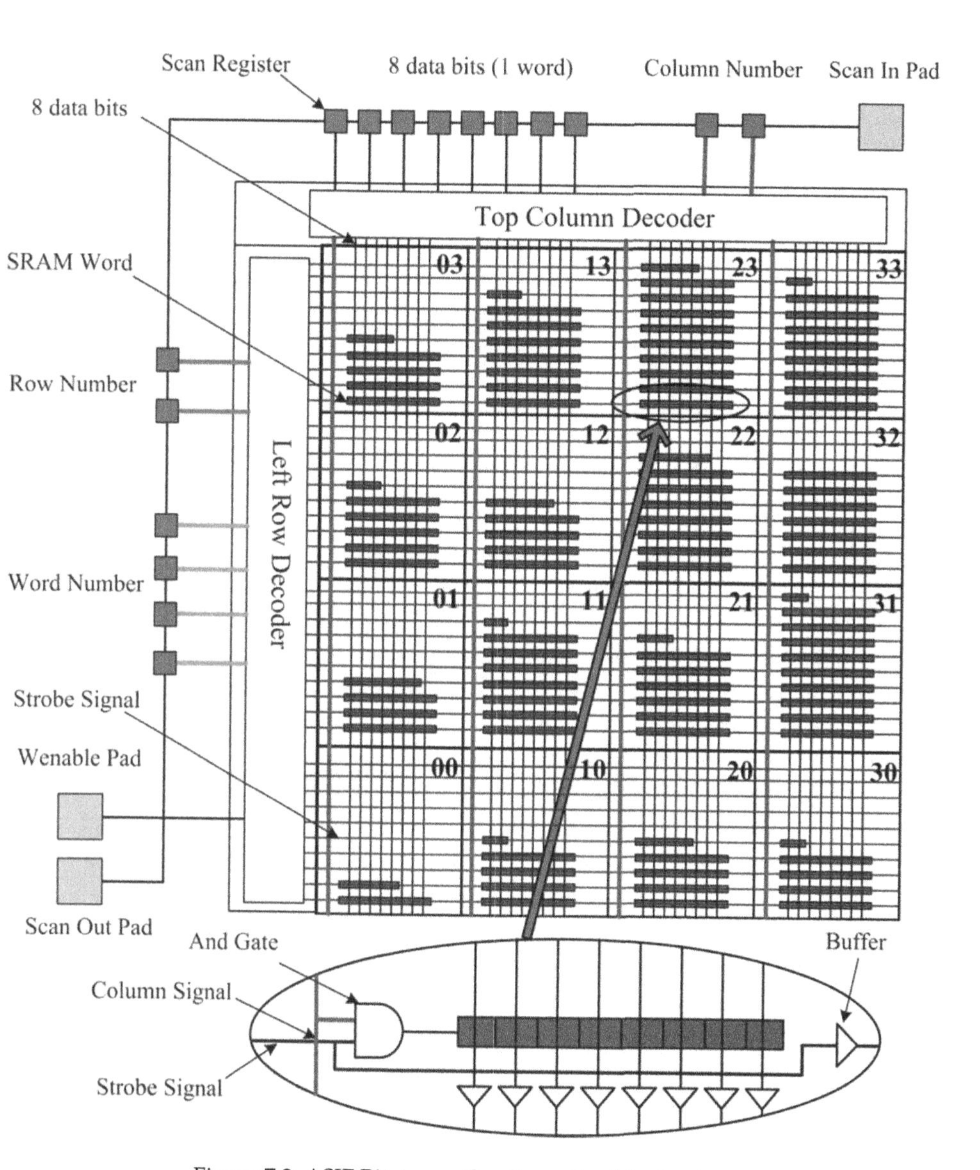

Figure 7.3: ASIF Bitstream Configuration Mechanism

selected SRAM word. The bitstream for each netlist contains the data bits to be written on a row number, column number and word number. One word is programmed at a time.

7.2.4 ASIF VHDL Generation

The VHDL model of an ASIF is generated by using ASIF floor-planning and ASIF routing graph. An ASIF architecture may contain different kinds of blocks including I/Os, CLBs and Hard-Blocks. The ASIF floor-planning gives the position of different blocks on the ASIF. These blocks are interconnected through a routing network. The routing network of the ASIF is represented by a routing graph. An ASIF routing graph contains nodes that are connected through edges; nodes represent a wire, and an edge represent the connections between different wires. A wire in the routing graph can be an input or output pin of a block, or a routing wire of the routing network.

An FPGA is transformed to an ASIF by reducing the routing graph of an FPGA. The reduced routing graph of an ASIF is later used to generate the VHDL model. The VHDL generation using routing graph is explained with the help of a small example as shown in Figure 7.4. Figure 7.4 (a) shows a unidirectional switch box of an FPGA having a channel width of 2. The routing graph for this switch box is shown in Figure 7.4 (b). This routing graph can be parsed to generate its physical representation. The physical representation of FPGA switch box is shown in Figure 7.4 (c). If a node is drived by more than one node, a multiplexor along with the required SRAMs is used to drive multiple nodes to the receiver node. If a node is drived by only a single node, a buffer is used to drive the receiver node. The physical representation of the routing graph is later translated to a VHDL model. This VHDL model is generated using a symbolic standard cell library, SXLIB [Alliance, 2006].

Figure 7.4 (d) and (g) show routing of two netlists on the FPGA switch box. The wires of the switch box used for routing these netlists are represented as thick/blue lines. The routing graph and physical representation of the switch box, for two routed netlists, are shown in Figure 7.4 (e), (f) and Figure 7.4 (h), (i) respectively. An ASIF switch box for the two netlists is shown in Figure 7.4 (j). The ASIF switch box only contains those switches and wires that are used by netlist-1 or netlist-2, the remaining switches and wires are removed. The routing graph for ASIF switch box is shown in Figure 7.4 (k). It can be seen that the ASIF routing graph is the sum of the routing paths of all the netlists. The physical representation of ASIF switch box, shown in Figure 7.4 (l), is generated by parsing the routing graph of ASIF.

The bitstream configuration mechanism and the layout of the ASIF is considered while generating the ASIF VHDL. Hence, before the transformation of an FPGA to an ASIF, layout related information is assigned to each routing wire. This is meant to determine the position of different physical components attached to routing wires. A routing wire, represented as a node in the routing graph, is assigned a tile number according to the position of wire on the FPGA as shown in Figure 7.2. The Figure shows the FPGA divided into different tiles Each tile is assigned a tile number, shown on the bottom-left of each tile. This tile number is assigned to each routing wire found in a tile. The routing wire can belong to a switch

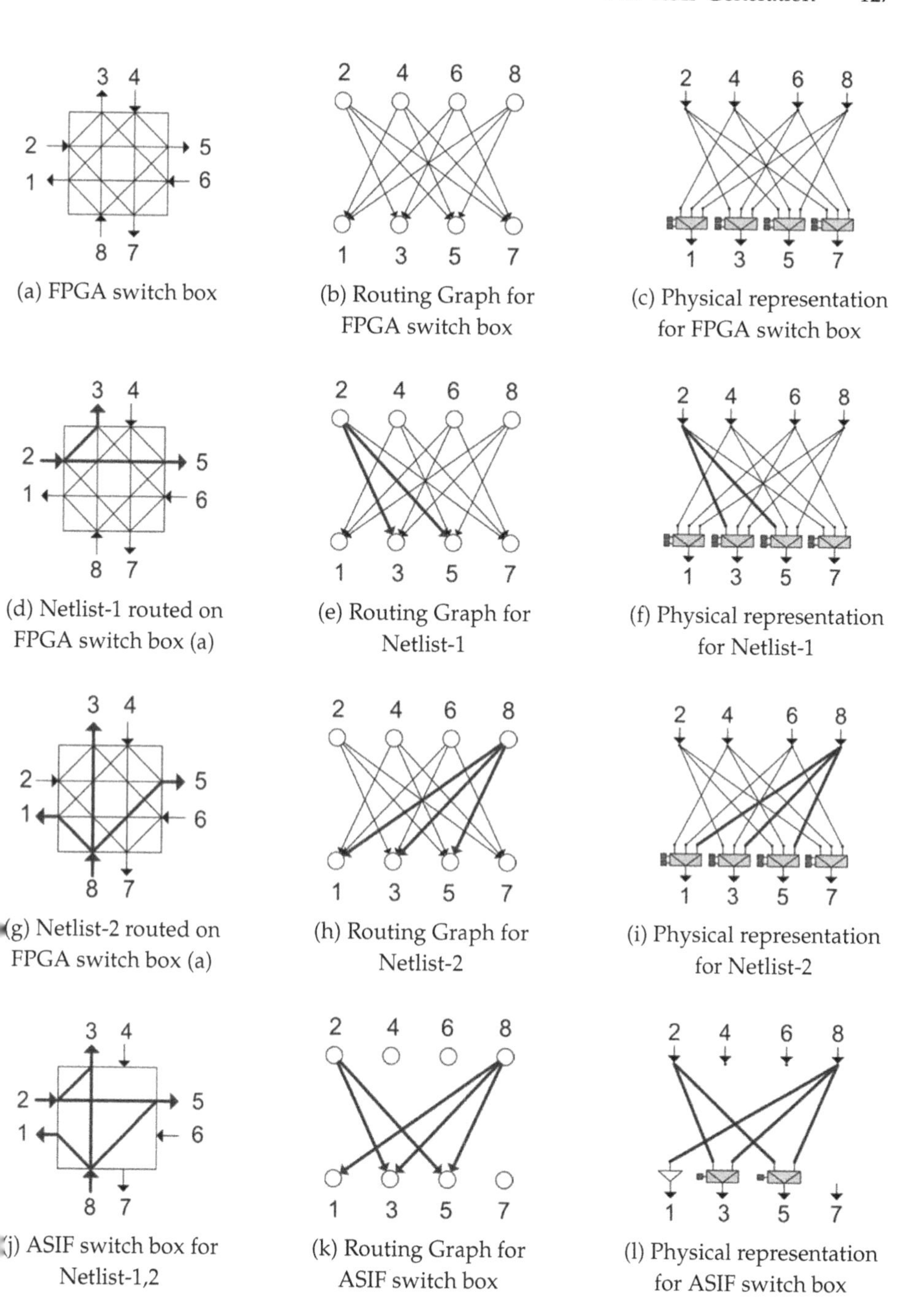

(a) FPGA switch box

(b) Routing Graph for FPGA switch box

(c) Physical representation for FPGA switch box

(d) Netlist-1 routed on FPGA switch box (a)

(e) Routing Graph for Netlist-1

(f) Physical representation for Netlist-1

(g) Netlist-2 routed on FPGA switch box (a)

(h) Routing Graph for Netlist-2

(i) Physical representation for Netlist-2

(j) ASIF switch box for Netlist-1,2

(k) Routing Graph for ASIF switch box

(l) Physical representation for ASIF switch box

Figure 7.4: ASIF VHDL generation from routing graph

box, a connection box, or input/output of a logic or IO block. After assigning tile numbers to routing wires, the entire FPGA architecture is transformed to routing graph. The FPGA routing graph is then reduced to an ASIF as explained in Figure 7.4. ASIF routing graph is then transformed to its physical representation. The VHDL generator will use the tile number of ASIF routing wires to determine the tile in which physical components of ASIF are placed.

The ASIF routing graph is parsed to generate its physical representation. If a receiver wire is drived by two or more wires, a multiplexor of appropriate size is connected to the receiver wire. If the receiver wire is being drived by only one wire, a buffer is used to connect the driver wire to the receiver wire. When a multiplexor is inserted, the SRAM bits required by the multiplexor are also declared along. The multiplexors and buffers belong to the same tile to which the receiver wire belongs. The SRAM bits connected to the multiplexor also belong to the same tile to which the multiplexor belongs. The total number of SRAMs in each tile are counted. Each time an SRAM is declared, the SRAM count of the particular tile is incremented. The SRAM count in a tile are used to determine the data and strobe signals which are to be connected to the SRAM. When the total data signals entering a tile are 8 (as shown in Figure 7.3), then the first SRAM bit declared in a tile will connect to first strobe and first data signal, the second SRAM bit will connect to first strobe and second data signal, the ninth SRAM bit will connect to second strobe and first data signal, and so on.

The IO and logic block instances are also declared. The logic block can be a soft-block such as a CLB, or a hard-block such as multiplier or adder etc. The input and output pins of these blocks are already represented in the routing graph. Thus, these blocks are automatically linked to physical representation of an ASIF. These blocks are declared in their respective tiles. The VHDL model of these blocks is provided along with the architecture description. The SRAMs used by any logic blocks are also placed in the same way as the SRAMs of routing network are placed.

The strobe and data signals required to program SRAMs are not included in the routing graph. Necessary buffers and AND gates, as shown in Figure 7.3, are inserted to column strobe and data signals.

7.2.5 Loader VHDL Generation

VHDL description of loader is generated along with the ASIF VHDL. A scan path is used to give configuration bitstream information to the loader. The loader then programs the bitstream on ASIF SRAMs. The number of scan registers in the scan path depend on the total ASIF rows, ASIF columns, maximum SRAM words in any ASIF tile, and data bits (which is equal to SRAM word size). The scan registers representing the data bits are propagated to SRAM words of an ASIF (from top to bottom as shown in Figure 7.3). The scan registers representing row, column and word number are used by row, column and word decoders. The number of outputs of row, column and word decoders represent the total number of ASIF rows, ASIF columns and maximum SRAM words in any ASIF tile. For the ASIF shown n Figure 7.3, the row and column decoders give 4 outputs, whereas the word decoder gives 10 outputs. Only one output of these decoders is set to high. The four outputs of column decoder are propagated to each column of the ASIF (the vertical red lines in Figure 7.3). The output of row decoder and word decoder are used to propagate the horizontal strobe signals (the horizontal blue lines in Figure 7.3). The 1-10 strobes entering the 1st row of ASIF iles are derived by anding the 1st output of row decoder, 1-10 outputs of word decoder and Wenable. The 1-10 strobes of 2nd row of ASIF tiles are derived by anding the 2nd output of row decoder, 1-10 outputs of word decoder and Wenable, and so on. The strobe signals entering a tile are anded with the column signal. This anded output is used by all SRAMs n an SRAM word. When the Wenable signal turns high, the 8 data bits are written to the elected SRAM word.

.2.6 Netlist Bitstream Generation

he netlist bitstream generator is used to generate the configuration bitstream for each netlist hat is mapped on ASIF. A netlist bitstream can be configured on the ASIF through the itstream loader shown in Figure 7.3. The netlist bitstream generator reads the ASIF floor-lanning and ASIF routing graph to determine the internal architecture of ASIF. The netlist outing information is parsed to know the exact wires and multiplexors used for routing the etlist on the ASIF. The wires used by a netlist give the value of SRAMs that control the mul-plexors driving these wires. The multiplexors of an ASIF that are not used by a particular etlists are initialized to 0.

he netlist function file contains LUT configuration values for all the LUTs used by the etlist. The netlist placement files and netlist function files are used to assign LUT config-ration values to the corresponding SRAMs in a tile. However, LUT configuration need to e modified in some cases, like when the nets are routed to LUT pins in a different order. can seen in Figure 7.2 that four inputs of a LUT-4 are connected to four adjacent routing hannels. Routing congestion can be slightly decreased if a net or signal is allowed to route the nearest LUT input, rather than to the exact LUT input as defined in the netlist file. The xploration environment allows the input pins of a CLB block to have same CLASS numbers. net or signal can be routed to any input pin of a block having same CLASS number. If the put pins of an instance are routed to LUT inputs in a different order than the original order a netlist file, the SRAM configuration of a Look-Up Table need to be changed accordingly.

uppose, a 4-input boolean function (F) needs to be mapped on a LUT-4, as shown in Fig-re 7.5. The Figure shows two LUTs that implement the boolean function (F). The input gnals of the boolean function are mapped to these two LUTs in different ordering; for first UT the order is ABCD, for second LUT the order is DBCA (from rightmost pin to leftmost in of LUT). The original configuration of the LUT is generated through SIS [E.M.Sentovich al., 1992] (as explained in section 3.4). The LUT configuration needs to be shuffled if the puts of the boolean function (F) are connected to the LUT inputs in different order. An gorithm is required that can automatically change the LUT SRAM values according to a different pin ordering. Equation 7.1 shows a rough algorithm for changing the LUT-4 configuration of the boolean function (F) when LUT pin connectivity changes from ABCD to DBCA. The algorithm shows that the SRAMs in a LUT are considered as an array. A new LUT configuration is computed from the original LUT configuration by simply swapping the SRAM values according to different pin orderings.

```
Loop A=0 to 1
  Loop B=0 to 1
    Loop C=0 to 1
      Loop D=0 to 1
        NewLUT [Dx8 + Bx4 + Cx2 + A] = OldLUT [Ax8 + Bx4 + Cx2 + D]        (7.1)
```

– When order of pin connectivity with LUT-4 changes from ABCD to DBCA
– NewLUT and OldLUT have 2^4 elements

The algorithm shown in Equation 7.1 shows a LUT configuration swapping for LUT-4 when pin order changes from ABCD to DBCA. There can be 16 different orderings of pins for a LUT-4. Besides, the ASIF exploration environment supports different LUT sizes. A generic algorithm needs to be written that can automatically change the configuration information of any LUT size using any pin ordering. Such a generic algorithm is shown in Figure 7.6. Line 02 of the Figure computes the new pin ordering. It compares the original pin ordering (as found in the netlist) and the routed pin ordering (as found in the netlist routing file). If the pins are connected in the same ordering, the pinOrderInfo is initialized to whole numbers (from 0 to N-1, where N is the total number of input pins of LUT). Considering the example shown in Figure 7.5, the original pin ordering is 0,1,2,3 (for ABCD) and new pin ordering is 3,1,2,0 (for DBCA). Line 04 calls a RecursiveLoop to compute the new LUT configuration for the new pin order.

7.2.7 Stimuli Generation

A test bench file and stimuli files are also generated for the simulation of ASIF. The test bench file connects the ASIF module with the stimuli module. A stimuli file reads the bitstream configuration of a netlist and programs it on the ASIF, word by word. Once the configuration file is programmed on ASIF, it reads the test vectors and gives them as input to the ASIF, and receives the output from ASIF. Later it compares the ASIF output with the results found in the test vectors. The test vectors are provided separately for each netlist.

The stimuli file reads the test vector inputs anc then connects them to the ASIF I/O blocks. This connection should be done according to the placement of I/O instances of the netlist on the I/O blocks of ASIF. Since the I/O instance placement for different netlists can be

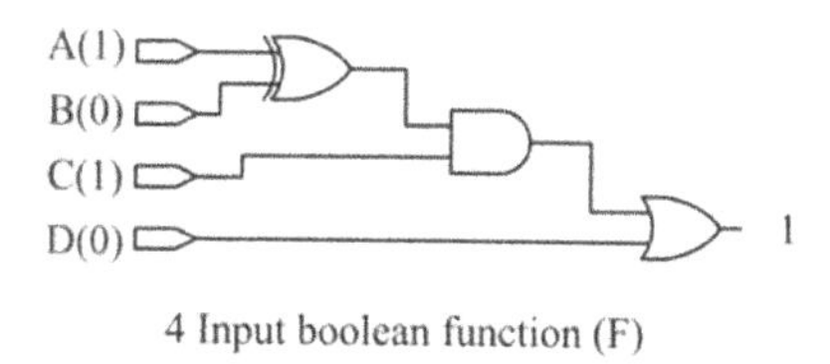

4 Input boolean function (F)

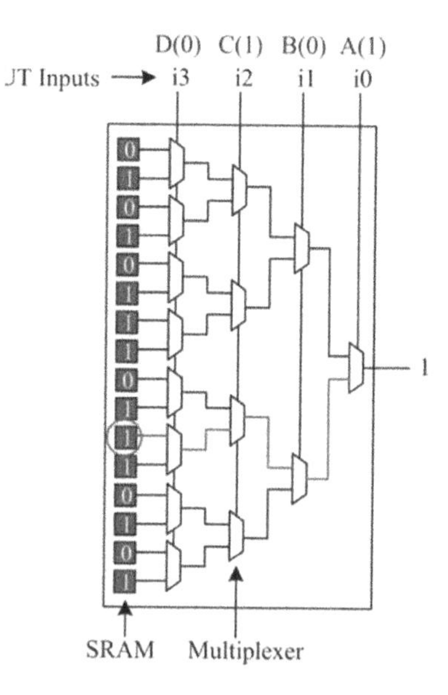

Look-Up Table for boolean function (F) with original pin connectivity

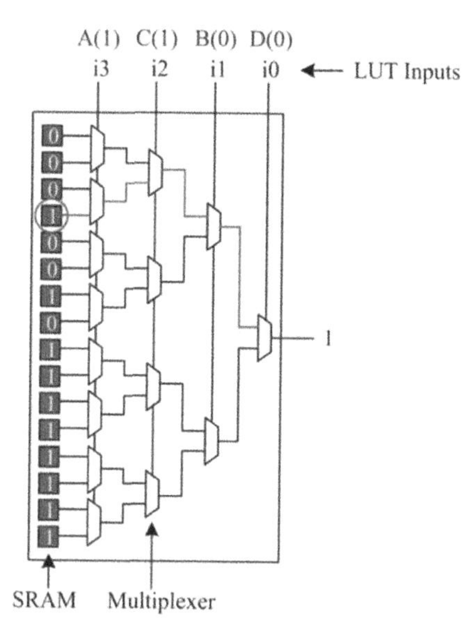

Look-Up Table for boolean function (F) with modified pin connectivity

Figure 7.5: Four input Look-Up Table configuration for different pin connectivity

different, a stimuli file is generated for each netlist file mapped on the ASIF. The stimuli files for different netlists differ only in their connection pattern of test vector IO with ASIF I/O.

7.3 Validation

ASIFs are generated for different number of netlists containing I/Os, CLBs and hard-blocks. Each ASIF is simulated using Synopsys. A netlist is individually programmed on the ASIF and their functionality is tested against the corresponding test vectors of the netlists.

The VHDL model of ASIF, loader, stimuli and testbench are simulated together. During the simulation, the stimuli reads the netlist configuration information and programs it on the ASIF. The netlist configuration contains SRAM data words and their addresses (i.e. column, row and word numbers). The SRAMs are programmed on the ASIF word by word. Once an

```
00  Foreach CLB instances {
01     ConfigInfo = Original configuration information for a LUT instance
02     pinOrderInfo = Compute new pin ordering (default pin ordering is 0,1,2,3...)
03     TotalInputPins = Total input pins of a LUT
04     RecursiveLoop (0, TotalInputPins)
05     ConfigInfo = newConfigInfo
06  }
07
08  RecursiveLoop (bit, pinNum) {
09     pinIndex[pinNum] = bit * (2 ^ pinNum)
10     newPinIndex[pinNum] = bit * (2 ^ pinOrderInfo[pinNum])
11
12     if(pinNum == 0) {
13        index = sum of all entries in pinIndex
14        newIndex = sum of all entries in newPinIndex
15        newConfigInfo[newIndex] = ConfigInfo[index]
16        return;
17     }
18
19     for(bit=0; bit<2; bit++)
20        RecursiveLoop ( bit, pinNum - 1)
21  }
```

Figure 7.6: Algorithm to change LUT configuration for different pin connectivity order

ASIF is programmed, it is ready to execute the functionality of the programmed application netlist. The stimuli reads the test vector and gives as input to the ASIF input blocks, and awaits the outputs of ASIF. The output of ASIF is then compared with the expected results.

Various tests have been performed successfully. These tests include different ASIF sizes, different heterogeneous blocks, different orientation of heterogeneous blocks, ASIFs for varying number of netlists, and with same CLASS numbers for LUT inputs (that eventually requires changes in LUT configuration).

7.4 ASIF Layout Generation

The ASIF VHDL model is generated using a symbolic standard cell library, SXLIB [Alliance, 2006]. The generated VHDL is translated to 130nm standard cell library of STMicroelectronics, and then passed to Cadence encounter for layout generation.

The size of VHDL files can become very huge for large sized ASIFs with higher channel width. It becomes very time consuming, or in some cases impossible for the commercial tools to handle very huge files. So, the VHDL model is automatically divided into smaller groups. Each group can contain a certain number of ASIF tiles. Architecture description parameters are used to know the number of groups in which the ASIF VHDL should be divided. Equivalence verification is performed between a flat VHDL model of an ASIF, and the VHDL model of the ASIF divided into several groups.

The generated VHDL model is translated to 130nm standard cell library of STMicroelectronics. This translation is done through Synopsys synthesis tool, Design Vision. The ST standard cell library used for experimentation does not contain an SRAM cell, so a LATCH is used. The area of a LATCH is generally larger than that of an SRAM. During the translation to ST cell library, all the buffers are removed. They are later automatically added in the layout phase.

The synthesized netlist is passed to Cadence encounter for layout generation. The fanout loads are fixed through automatic buffer insertion and gate resizing. The default parameters of encounter enforce at least 5% space reserved for empty space (fillers). ASIF layout generation uses these default parameters. If there is too much congestion in the chip, more fillers can be inserted to facilitate routing of a chip. The ASIF sizes are selected in such a way that the desired amount of buffers are inserted and gate sizing is performed while maintaining the minimum 5% space for fillers. Several core sizes are tested to find the minimum routable core layout for an ASIF.

7.5 Experiments and Analysis

One of the major concerns from previous chapters is to find the maximum channel width for which ASIF layout is not wire dominant. Due to some preferences of efficient wire sharing, increasing the channel width of an FPGA decreases the total area of ASIF, however the number of wires used in the ASIF increase. If the layout becomes wire dominant, increasing the channel width will not be beneficial anymore. So a best compromise needs to be searched. Layout is performed, and their total area is measured for ASIFs with varying channel widths.

7.5.1 Opencores circuits

Different ASIFs are generated for a group of 10 opencore netlists as presented in Table 6.1 and 6.2. Layout is done only for LUT-3 version and gates version of opencore netlists. The LUT-3 version of opencore netlists are synthesized using LUT-3, multipliers and adders. The gates version of opencore netlists are synthesized using commonly used gates, multipliers and adders. The gates, LUT-3, multipliers and adders are based on standard cell library (SXLIB). An ASIC version of these netlists is also synthesized using any set of gates found in the standard cell library (SXLIB). The sum of ASICs of these opencore netlists is a single netlist which combines the ASIC version of all the netlists. VHDL is generated for LUT-3 version based and gates version based ASIFs for 10 opencore netlists, and they are then compared with sum of ASICs.

(a) ASIF LUT-3, cw-16, 2.62 mm^2

(b) ASIF LUT-3, cw-24, 2.37 mm^2

(c) ASIF LUT-3, cw-40, 2.17 mm^2

(d) ASIF LUT-3, cw-80, 2.15 mm^2

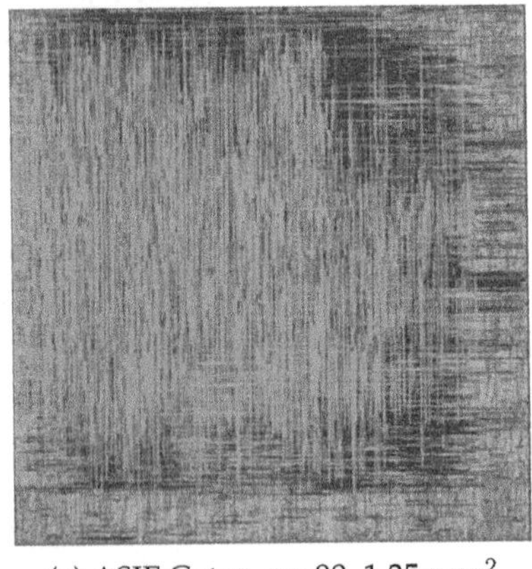

(e) ASIF Gates, cw-80, 1.35 mm^2

(f) Sum of ASICs, 1.25 mm^2

(Legend: Metal-6 orange, Metal-5 maroon, Metal-4 yellow, Metal-3 green, Metal-2 red, Metal-1 blu

Figure 7.7: ASIC/ASIF layout for 10 opencores circuits using 130nm process of ST

7.5.2 Results and Analysis

The VHDL model of LUT-3 and gates based ASIFs, and sum of ASICs is transformed to 130nm standard cell library of STMicroelectronics. Their layout is then performed using 130nm 6-metal layer CMOS process of ST. Figure 7.7 shows the core layout of sum of ASICs, and the ASIFs with varying channel widths. The color of routing wires represent the metal layer on which different wires are routed. A legend representing the color of different metal layers is shown at the bottom of Figure 7.7. Generally, lower metal layers are preferred for routing. However as routing congestion increases, upper metal layers are also used. It can be seen that as the channel width (cw) increases, the ASIF area decreases, whereas the routing congestion increases. However, the ASIF cores are routable with all different channel widths. The sum of ASICs has relatively less congestion than the ASIFs.

Buffer insertion and gate resizing is performed to fix fanout loads in the layouts. The ASIFs for varying channel widths require 1.7% to 3.5% area of the core for buffer insertion and gate resizing. ASIFs with less channel widths require less area for fixing fanout loads, and greater channel widths require more area for fixing fanout loads. The sum of ASICs use only 0.9% area of the chip for fixing fanout loads. The ASIFs and sum of ASICs use 5.2% to 5.8% area of the core for fillers.

The graph in Figure 7.8 shows the area of different ASIF layouts in millimeter square. The ST standard cell library does not contain SRAM cell. Thus a LATCH is used instead of SRAM. Generally a LATCH cell takes more area than an SRAM cell. Considerable amount of ASIF area increases due to the use of LATCH instead of SRAM. Figure 7.9 shows the percentage area taken by SRAMs in ASIFs. The percentage of SRAM area in ASIFs is measured from the VHDL model generated using the symbolic standard cell library SXLIB (i.e measured before mapping it on ST standard cell library). The SRAMs in an ASIF are either used in LUTs or by the multiplexors of routing network. The gates version of ASIF does not use LUTs, only their routing network requires SRAMs. That is why gates version require less SRAMs than LUT-3 version. The sum of ASICs does not use any SRAM. The SRAMs in LUT-3 based ASIF for different channel widths take 26.5% to 29.7% area of the chip. The SRAMs in gates based ASIF for different channel widths take 16.5% to 22.1% area of the chip. The high percentage of SRAMs used in ASIFs show that considerable amount of area can be reduced when LATCH cells are replaced by SRAM cells in the ST standard cell library based layout.

7.6 Conclusion and Future Work

This chapter has presented an automated method to generate hardware description for ASIFs. The hardware generator is directly integrated with the exploration environment, thus the VHDL model can be generated for any ASIF architecture supported by the exploration environment. Bitstreams are also generated for each netlist mapped on an ASIF. Bitstreams of different application circuits are programmed and executed on the ASIF at exclusive times.

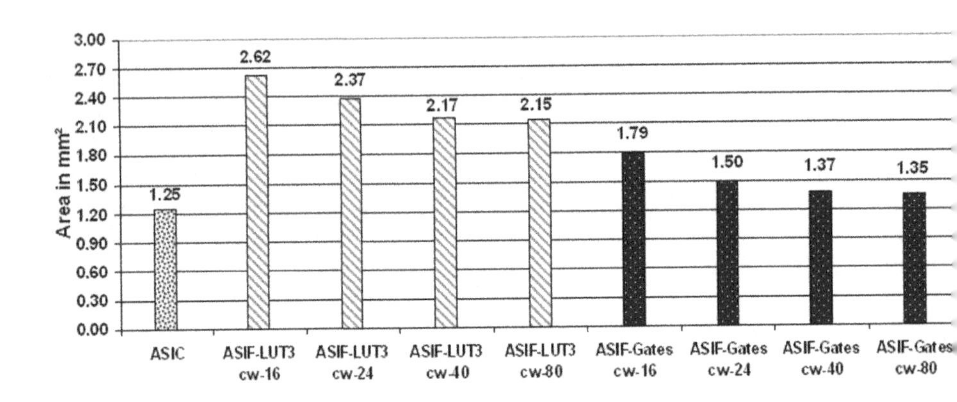

Figure 7.8: ASIC vs ASIF for 10 opencores circuits using 130nm ST standard cell library

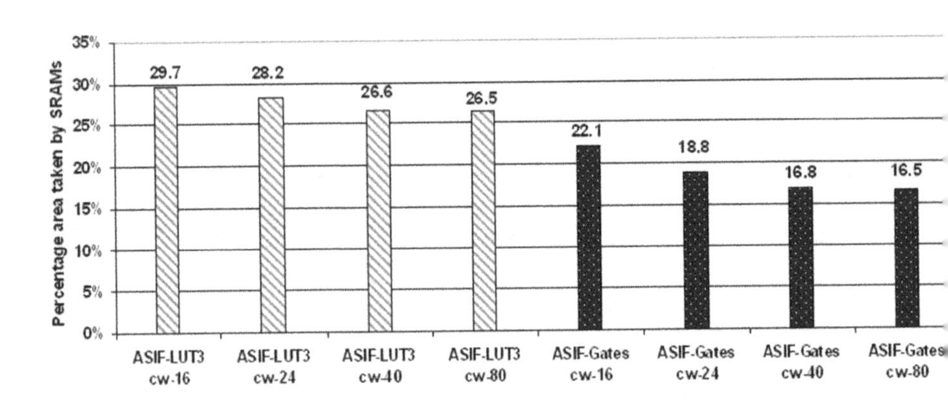

Figure 7.9: Percentage SRAM area in ASIFs (Layout done with LATCH instead of SRAM)

Layout is performed for ASIFs using 130nm 6-metal layer CMOS process of ST. ASIFs are routable with high channel widths, however their routing congestion increases. A LUT-3 based ASIF core for 10 opencore netlists, using channel width of 80, measures 2.15 mm^2. A gates based ASIF core for 10 opencore netlists, using channel width of 80, measures 1.35 mm^2. The sum of ASICs for 10 opencore netlists measures 1.25 mm^2. However, the area of ASIFs is greatly affected due to the unavailability of SRAM cell in ST standard cell library. Instead of SRAM, a LATCH is used which eventually increases the area of ASIFs.

In future, the SRAM cell can designed to replace LATCH cell. Moreover, full-custom designs of repeatedly used cells or blocks can be performed to further decrease the area of ASIFs. These full-custom attempts cannot be very beneficial for sum of ASICs, as there are no major repeatedly used components in ASICs. However, as an ASIF is reduced from an FPGA, a lot of commonly used components can be found. These repeatedly used components may include a group of 4 or 8 SRAMs, LUTs and multiplexors etc. Full-custom design of these group of components will further decrease area of an ASIF.

8

Conclusion and Future Lines of Research

8.1 Summary of contributions

This work has explored application specific mesh-based heterogeneous FPGA architectures and their automatic hardware generation. Following are few of the major contributions of this work.

8.1.1 Heterogeneous architecture exploration environment

Chapter 3 presented a new environment for the exploration of mesh-based heterogeneous FPGA architectures. One of the key features of this environment is its ability to optimize floor-planning of mesh-based heterogeneous FPGA architectures. When a domain-specific FPGA architecture is required for a known set of application designs, floor-planning can be automatically optimized for the given designs.

Commercial FPGA architectures commonly use embedded hard-blocks to fulfill varying domain-specific requirements of their customers. Floor-planning of these FPGAs is optimized by placing similar type of blocks in columns. It has been shown that the floor-planning achieved by moving and rotating the blocks is much better than column-based floor-planning. A column-based floor-planning requires on average 35% more placement cost, 24% more channel width, and thus 26% more area as compared to a floor-planning achieved by moving and rotating the blocks. For netlists having more than two types of hard-blocks, the area difference is as high as 55%. A column-based layout contains only a single type of block in each column. The number of hard-blocks in a column-based FPGA are thus restricted to

H. Parvez and H. Mehrez, *Application-Specific Mesh-based Heterogeneous FPGA Architectures*,
DOI 10.1007/978-1-4419-7928-5_8, © Springer Science+Business Media, LLC 2011

be in multiples of hard-blocks that can fit in a column. Some hard-blocks remain unused if netlist requirement does not exactly match the number of hard-blocks on the architecture. Moreover, communication between different hard-blocks, placed in distinct columns, require additional routing resources, thus FPGA area increases. These area losses will further deteriorate as more types of hard-blocks are required in an FPGA to efficiently support new domain-specific requirements of customers. However, the move and rotate floor-planning will lose some area benefits if a tile-based layout is required. These area advantages are retained if the VHDL model of the final FPGA is taped-out using any commercial ASIC design flow.

8.1.2 Tile-based layout generator for FPGAs

Chapter 4 presented an automated method for the generation of a tile-based FPGA layout. The main purpose of the generator is to reduce overall time-to-market of domain-specific FPGA architectures. The tile-based layout generator employs an elegant scheme to introduce manual intervention in FPGA layout generation. The manual intervention is achieved by generating the layout in two phases. In the first phase, partial layouts of different tiles are generated. The partial layout generator takes a number of different layout parameters to modify the layout of different tiles according to initial requirements. In the second phase, the remaining layout is completed by using automatic placement and routing tools. All the different tiles are abutted together to form a complete FPGA. This generator generates a technology independent layout that can be migrated to any process technology. The proposed method is validated by fabricating a 32x32 FPGA layout using 130nm process technology of ST.

8.1.3 ASIF : Application Specific Inflexible FPGA

Chapter 5 proposed Application Specific Inflexible FPGAs. If a digital product is required to provide multiple functionalities at exclusive times, each distinct functionality represented by an application circuit is efficiently mapped on an FPGA. Later, unused resources of the FPGA are removed to generate an ASIF. ASIF generation techniques can also be employed to generate a single configurable ASIC core that can perform multiple tasks at different times.

Different ASIF generation techniques were initially presented. ASIF-1 technique places the netlists normally; however no routing resources are shared between different netlists. This technique drastically increases the routing wires as number of netlist increases. The layout area of ASIF-1 for large number of netlists might eventually be dominated by routing wires. ASIF-2 places and routes the netlists normally. The number of routing wires decrease, but more switches are required. ASIF-3 places normally, but routing is done efficiently. ASIF-4 places and routes efficiently. The major benefit of ASIF-4 is its ability to give a compromised switch-count/wire-count solution that lies between ASIF1 and ASIF2.

In an FPGA, the logic area normally occupies 10 to 20% of the total area. The remaining 80 to 90% area is occupied by routing resources. However in an ASIF, the routing area is reduced to such an extent that logic area occupies a very major portion of an ASIF. ASIFs are generated with varying LUT sizes. For smaller LUT size, logic area reduces as LUT size is reduced. However, more LUTs are required and thus more routing resources are required to connect them. For MCNC benchmark circuits, an ASIF generated using LUT-2 gives minimum area results.

The quality analysis of ASIF generation method is performed by generating an ASIF for a group of netlists for which an ideal ASIF solution is known. So, an ASIF is generated for same set of netlists. It is found that an ASIF generated for 10 same netlists is almost 50% better than the worst-case ASIF solution, but 50% worst than the ideal solution. The major cause of area loss is the heuristic based efficient placement algorithm. New placement techniques need to be explored to further improve area of an ASIF.

Chapter 6 introduced Heterogeneous blocks in ASIFs. Their inclusion is meant to decrease the gap between ASIC and ASIFs. The position of different heterogeneous blocks is optimized to get maximum area gains. It has been shown that an ASIF for 10 opencores circuits can be less than the sum or areas of individual ASICs.

The logic resources are not fully optimized, reconfigurability in these logic resources can be exploited to map new or modified version of netlists on an ASIF. The normal heuristic based algorithms do not allow to find a solution from the solution space having only one or few solution. Chapter 6 also presented a new algorithm to map netlists on an ASIF.

8.1.4 ASIF Hardware Generation

Chapter 7 presented hardware layout generation for ASIF. The technique for ASIF generation is different from the tile-based FPGA layout generator. The hardware generator is integrated with the exploration environment. So the major benefit is that all the architecture level changes in the exploration environment are directly translated in VHDL model. The VHDL model of an ASIF is generated and simulated for multiple netlists. The bitstream configuration information is also generated for each netlist mapped on ASIF. The bitstream loader program the netlist bitstream on an ASIF. Bitstreams of different application circuits are programmed and executed on the ASIF at exclusive times.

Layout is performed for ASIFs using 130nm 6-metal layer CMOS process of ST. ASIFs are routable with high channel widths, however their routing congestion increases. The LUT-3 based ASIF core for 10 opencore netlists measures 2.15 mm^2. A gates based ASIF core for 10 opencore netlists measures 1.35 mm^2. The sum of ASICs for 10 opencore netlists measures 1.25 mm^2. However, the area of ASIFs is greatly affected due to the unavailability of SRAM cell in ST standard cell library. Instead of SRAM, a LATCH is used which eventually increases the area of ASIFs.

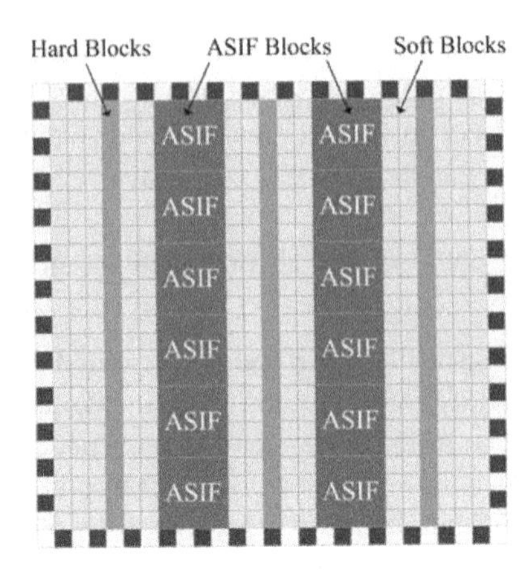

Figure 8.1: An illustration of ASIF blocks integrated in an FPGA

8.2 Future work

The future work includes the following main directions

- **Integrating ASIF blocks in an FPGA/ASIC**

  Commercial FPGA vendors use different variety of hard-blocks in their FPGAs. They provide a range of FPGA device variants to fulfill varying domain-specific requirements of their customers. Smaller ASIF blocks can also be integrated in FPGAs to enhance the domain-specific needs of customers. An ASIF block serves as a multi-tasking hard-block that can support a number of different functionalities at mutually exclusive times. An illustration of an FPGA with ASIF blocks is shown in Figure 8.1. In order to get maximum advantage of an ASIF block, the total area of an ASIF block should be less than the area taken by any of the ASIF functionality if it was directly implemented on FPGA fine-grain resources. ASIF blocks can be designed for some general purpose DSP requirements, or for more specific applications such as video decoders/encoders applications etc.

  Similarly, ASIF blocks can also be used in ASICs to incorporate limited and confined reconfigurability in ASIC solutions.

- **Area improvement of an ASIF**

  The quality measurement of an ASIF (as presented in section 5.7) shows that the area of an ASIF can still be improved. The major area loss in an ASIF is due to heuristic-based placement algorithm. The existing algorithms need to be improved/fine-tuned to achieve maximum possible gains. New placement solutions can be proposed to attain better solution in limited time period.

  Another dimension to improve the area of an ASIF is to generate ASIF from different FPGA architectures. The traditional mesh-based FPGA architecture, used in this work provide many placement and routing possibilities. Too much flexibility in the FPGA architectures limit the heuristic based algorithms to converge the commonalities of different netlists. Due to this reason, less flexible FPGA architectures might render better area improvements when reduced to an ASIF. We believe that an ASIF for a tree-based FPGA architecture [Marrakchi, 2008] might give far better results than a mesh-based FPGA.

- **SRAM cell layout, Full-Custom layout to reduce ASIF area**

  In this work, ASIF layout is performed using LATCH cells instead of SRAM cells. In future, the area of ASIF layout can be reduced by designing SRAM cells. Moreover, full-custom designs of repeatedly used cells or blocks can be performed to further decrease the area of ASIFs. A lot of commonly used components can be found in an ASIF. These repeatedly used components may include a group of 4 or 8 SRAMs, LUTs and multiplexors etc. Full-custom design of these group of components will further decrease area of an ASIF.

- **Timing analysis of an ASIF**

  One of the major future works is to perform timing analysis of an ASIF. The timing analysis will include measuring the critical path of an ASIF, optimizing its critical path, and finding a good compromised ASIF solution for timing and area benefits.

- **Optimizing logic resources in an ASIF**

  In a traditional FPGA architecture, logic blocks resources occupy only 10 to 20% area of an FPGA. The remaining 80 to 90% of FPGA area is occupied by routing resources. Since an ASIF is generated by removing unused routing resources of an FPGA, the logic area percentage in an ASIF is higher than an FPGA. In a LUT-2 based ASIF for 17 MCNC netlists, 32% area is occupied by logic resources. For larger LUT sizes, logic area percentage is much higher. In a LUT-4 based ASIF for 17 MCNC netlists, 42% area is occupied by logic resources. An ASIF for lesser number of netlists requires lesser routing resources. Thus, logic area percentage further increases as the number of netlists decrease. In this context, one of the key future direction is the optimization of logic block resources in an ASIF.

  Different set of application netlists, mapped on an ASIF, program the SRAM bits of a LUT differently. If all the netlists program a particular SRAM of a LUT in a similar fashion, that SRAM bit can be replaced by a hard-coded 0 or 1. Similarly, if a 2-input multiplexor in a LUT receives similar hard-coded values, that multiplexor can be removed, and replaced by the hard-coded value. A major element of research is to tailor the LUT configuration of different netlists in such a way that maximum logic block optimization is achieved. One possible option is to exploit the don't care input pins of LUTs of different netlists. The SRAM bits effected by don't care inputs can be modified to maximize common SRAM values for all the netlists.

  Another option is to modify the order of pins of LUT instances. Changing the order of pins requires to change SRAM configruation of LUT (as explained in section 7.2.6). After the placement phase, the best pin ordering of LUTs can be searched for each LUT instance of each netlist so that maximum common SRAMs can be found. The pin ordering can then be fixed before the routing phase. The major challenge in logic block optimization is to develop new algorithms to perform these optimizations in a reasonable time. Logic block optimization can give area benefits in an ASIF for less number of netlists.

- **Improved mapping of new application designs on an ASIF**

  Placement algorithms need to be improved to efficiently map new application circuits on an ASIF. Target netlist modifications can be performed to map new netlists on an ASIF. The netlist modifications may include addition of buffers and instance replication to better adapt the new netlist according to ASIF characteristics. Besides, reconfigurable resources in an ASIF can be intelligently increased so that more new applications can be easily mapped on an ASIF.

Bibliography

[A.DeHon, 1999] A.DeHon (1999). Balancing Interconnect and Computation in a Reconfigurable Computing Array (or, why you don't really want 100% LUT utilization). *International symposium on Field Programmable Array FPGA, Monterey, CA*, pages 69 – 78.

[A.Dunlop and B.Kernighan, 1985] A.Dunlop and B.Kernighan (1985). A Procedure for Placement of Standard-cell VLSI Circuits. *IEEE Transactions on CAD*, pages 92–98.

[Alliance, 2006] Alliance (2006). *http://www-asim.lip6.fr/recherche/alliance/.*

[Altera, 2010] Altera (2010). *http://www.altera.com.*

[A.Marquart et al., 1999] A.Marquart, V.Betz, and J.Rose (1999). Using Cluster-based Logic Block and Timing-driven Packing to improve FPGA Speed and Density. *International symposium on FPGA, Monterey*, pages 37–46.

[ATMEL, 2010] ATMEL (2010). *http://www.atmel.com.*

[Beauchamp et al., 2006] Beauchamp, M., Hauck, S., Underwood, K., and Hemmert, K. (2006). Embedded floating-point units in FPGAs. *FPGA*, pages 12–20.

[Beilleau et al., 2009] Beilleau, N., Aboushady, H., Montaudon, F., and Cathelin, A. (2009). A 1.3V 26mW 3.2GS/s Undersampled LC Bandpass Sigma-Delta ADC for a SDR ISM-band Receiver in 130nm CMOS. *IEEE Radio Frequency Integrated Circuits Symposium, RFIC'09*, pages 383–386.

[Belloeil et al., 2007] Belloeil, S., Dupuis, D., Masson, C., Chaput, J., and Mehrez, H. (2007). Stratus: A procedural description language based upon Python. *19th International Conference on Microelectronics*, pages 275–278.

[Berkeley Logic Synthesis and Verification Group, 2005] Berkeley Logic Synthesis and Verification Group (2005). ABC: A System for Sequential Synthesis and Verification. *http://www.eecs.berkeley.edu/ alanmi/abc/.*

[Betz et al., 1999] Betz, V., Marquardt, A., and Rose, J. (1999). Architecture and CAD for Deep-Submicron FPGAs. *Kluwer Academic Publishers.*

[Brayton et al., 1990] Brayton, R., G.Hachtel, and Sangiovanni-Vincentelli, A. (1990). Multilevel logic synthesis. *Proceedings of the IEEE*, 78(2):264–300.

H. Parvez and H. Mehrez, *Application-Specific Mesh-based Heterogeneous FPGA Architectures*,
DOI 10.1007/978-1-4419-7928-5,

[Brayton and McMullen, 1982] Brayton, R. and McMullen, C. (1982). The decomposition and factorization of Boolean expressions. *Proc. ISCAS*, pages 29–54.

[Cadence, 2010] Cadence (2010). Cadence. *http://www.cadence.com.*

[Callahan et al., 2000] Callahan, T. J., Hauser, J. R., and Wawrzynek, J. (2000). The Garp architecture and C compiler. *Computer*, 33(4):62–69.

[Cherepacha and Lewis, 1996] Cherepacha, D. and Lewis, D. (1996). DP-FPGA: An FPGA Architecture Optimized for Datapaths. *VLSI Design*, 4(4):329–343.

[C.H.Ho et al., 2006] C.H.Ho, P.H.W.Leong, W.Luk, S.Wilton, and S.Lopez-Buedo (2006). Virtual Embedded Blocks: A Methodology for Evaluating Embedded Elements in FPGAs. *FCMM*, pages 35–44.

[C.J.Alpert et al., 1997] C.J.Alpert, T.Chan, D.Huang, A.Kahng, I.Markov, P.Mulet, and K.Yan (1997). Faster Minimization of Linear Wirelength for Global Placement. *ACM Symposium on Physical Design*, pages 4–11.

[Compton and Hauck, 2007] Compton, K. and Hauck, S. (2007). Automatic Design of Area-Efficient Configurable ASIC Cores. *IEEE Transaction on Computers*, 56(5):662–672.

[C.Sechen and A.Sangiovanni-Vincentelli, 1985] C.Sechen and A.Sangiovanni-Vincentelli (1985). The Timberwolf Placement and Routing Package. *JSSC*, pages 510–522.

[D.Huang and A.Kahng, 1997] D.Huang and A.Kahng (1997). Partitioning-based Standard-cell Global Placement with an Exact Objective. *ACM Symposium on Physical Design*, pages 18–25.

[E.Ahmed and J.Rose, 2000] E.Ahmed and J.Rose (2000). The Effect of LUT and Cluster Size on Deep-submicron FPGA Performance and Density. *Proceedings of the International Symposium on Field Programmable Gate Arrays*, pages 3–12.

[eASIC, 2010] eASIC (2010). *http://www.easic.com.*

[Ebeling, 2002] Ebeling, C. (2002). Rapid-C Manual. *University of Washington Technical Report : UW-CSE-02-07-06.*

[Ebeling et al., 1996] Ebeling, C., Cronquist, D. C., and Franklin, P. (1996). RaPiD - Reconfigurable Pipelined Datapath. *Field Programmable Logic and Applications*, pages 126–135.

[Elder et al., 1988] Elder, J., Osborn, J., Kolasinski, W., and Koga, R. (1988). A method for characterizing a microprocessor's vulnerability to SEU. *IEEE Transaction on Nuclear Science*, 35(6).

[E.M.Sentovich et al., 1992] E.M.Sentovich, K.J.Singh, L.Lavagno, C.Moon, R.Murgai, A.Saldanha, H.Savoj, Stephan, P., R.K.Brayton, and A.Sangiovanni-Vincentelli (1992). SIS: A System for Sequential Circuit Synthesis. *Technical Report No. UCB/ERL M92/41. University of California, Berkeley.*

[Essen et al., 2009] Essen, B. V., Wood, A., Carroll, A., Friedman, S., Panda, R., Ylvisaker, B., Ebeling, C., and Hauck, S. (2009). Static Versus Scheduled Interconnect in Coarse-Grained Reconfigurable Arrays. *International Conference on Field Programmable Logic and Applications*, pages 268–275.

[Friedman, 2001] Friedman, E. G. (2001). Clock Distribution Networks in Synchronous Digital Integrated Circuits. *Proceedings of the IEEE*, pages 665–692.

[GAUT, 2010] GAUT (2010). High-Level Synthesis tool From C to RTL. *http://www-labsticc.univ-ubs.fr/www-gaut/.*

[G.Lemieux et al., 2004] G.Lemieux, E.Lee, M.Tom, and A.Yu (2004). Directional and Single-Driver Wires in FPGA Interconnect. *IEEE International Conference on Field-Programmable Technology (ICFPT)*, pages 41–48.

[G.Sigl et al., 1991] G.Sigl, K.Doll, and F.Johannes (1991). Analytical Placement: A Linear or a Quadratic Objective Function? *Design Automation Conference*, pages 427–432.

[HardCopy, IV] HardCopy (IV). HardCopy IV ASICs, Device Handbook. *Available at http://www.altera.com/products/devices/hardcopy-asics/hardcopy-iv/literature/hciv-literature.jsp.*

[Hartenstein, 2001] Hartenstein, R. (2001). Coarse Grain Reconfigurable Architectures (embedded tutorial). *Asia and South Pacific Design Automation Conference*, pages 564–570.

[Hutton et al., 2006] Hutton, M., Yuan, R., Schleicher, J., Baeckler, G., Cheung, S., Chua, K., and Phoon, H. (2006). A Methodology for FPGA to Structured-ASIC Synthesis and Verification. *DATE*, 2:64–69.

[I.Kuon and J.Rose, 2007] I.Kuon and J.Rose (2007). Measuring the Gap Between FPGAs and ASICs. *IEEE Transactions on CAD*, 26(2):203–215.

[Jamieson and J.Rose, 2006] Jamieson, P. and J.Rose (2006). Enhancing the Area-Efficiency of FPGAs with Hard Circuits Using Shadow Clusters. *IEEE International Conference on Field-Programmable Technology (ICFPT)*, pages 1–8.

[J.Cong and Y.Ding, 1994a] J.Cong and Y.Ding (1994a). FlowMap: An Optimal Technology Mapping Algorithm for Delay Optimization in Lookup-Table based FPGA Designs. *IEEE Transactions on Computer-Aided Design*, pages 1–12.

[J.Cong and Y.Ding, 1994b] J.Cong and Y.Ding (1994b). On Area/Depth Trade-off in LUT-Based FPGA Technology Mapping. *IEEE Transactions on VLSI Systems*, 2(2):137–148.

[J.Cong and Y.Ding, 2000] J.Cong and Y.Ding (2000). Structural Gate Decomposition for Depth-Optimal Technology in LUT-Based FPGA Designs. *ACM Transactions on Design Automation of Electronic Systems*, 5(3).

[J.Cong and Y.Hwang, 1995] J.Cong and Y.Hwang (1995). Simultaneous Depth and Area Minimization in LUT-Based FPGA Mapping. *ACM/SIGDA International Symposium on Field Programmable Gate Array*, pages 68–74.

[J.Greene et al., 1993] J.Greene, E.Hamdy, and S.Beal (1993). Antifuse Field Programmable Gate Arrays. *Proceedings of the IEEE*, pages 1042–1056.

[Jones et al., 2005] Jones, A. K., Hoare, R., Kusic, D., Fazekas, J., and Foster, J. (2005). An FPGA-based VLIW Processor with Custom Hardware Execution. *Proceedings of the International Symposium on Field Programmable Gate Arrays*, pages 107–117.

[J.Rose et al., 1990] J.Rose, R.Francis, D.Lewis, and P.Chow (1990). Architecture of Field-Programmable Gate Arrays: The Effect of Logic Functionality on Area Efficiency. *IEEE Journal of Solid State Circuits*, 25(5):1217 – 1225.

[Kapre et al., 2006] Kapre, N., Mehta, N., deLorimier, M., Rubin, R., Barnor, H., J.Wilson, M., Wrighton, M., and DeHon, A. (2006). Packet Switched vs. Time Multiplexed FPGA. *IEEE Symposium on Field-Programmable Custom Computing Machines (FCCM)*, pages 205–216.

[Kirkpatrick et al., 1983] Kirkpatrick, Gelatt, and Hecchi (1983). Optimisation by Simulated Annealing. *Science*, 220(4598):671–680.

[Kuon et al., 2005] Kuon, I., Egier, A., and Rose, J. (2005). Design, layout and verification of an FPGA using automated tools. *FPGA*, pages 215–226.

[Lagadec, 2000] Lagadec, L. (2000). Abstraction, modélisation et outils de CAO pour les architectures reconfigurables. *PhD. Thesis : Université de Rennes 1.*

[Lawler and Wood, 1966] Lawler, E. L. and Wood, D. E. (1966). Branch-and-bound methods: A survey. *Operations Research*, 14:699–719.

[Lemieux et al., 2004] Lemieux, G., Lee, E., Tom, M., and Yu, A. (2004). Directional and Single-Driver Wires in FPGA Interconnect. *IEEE International Conference on Field-Programmable Technology (ICFPT)*, pages 41–48.

[L.McMurchie and C.Ebeling, 1995] L.McMurchie and C.Ebeling (1995). Pathfinder: A Negotiation-Based Performance-Driven Router for FPGAs. *International Workshop on Field Programmable Gate Array*, pages 111 – 117.

[Luu et al., 2009] Luu, J., Kuon, I., Jamieson, P., Campbell, T., Ye, A., Fang, W. M., and Rose, J. (2009). VPR 5.0: FPGA CAD and Architecture Exploration Tools with Single-Driver Routing, Heterogeneity and Process Scaling. *FPGA*, pages 133–142.

[Marrakchi, 2008] Marrakchi, Z. (2008). Exploration and Optimization of Tree-based FPGA Architectures. *PhD. Thesis : http://www-asim.lip6.fr/publications/.*

[Marrakchi et al., 2009] Marrakchi, Z., Mrabet, H., Farooq, U., and Mehrez, H. (2009). FPGA Interconnect Topologies Exploration. *Hindawi Publishing Corporation*, 2009(2598):1–13.

[Marshall et al., 1999] Marshall, A., Stansfield, T., Kostarnov, I., Vuillemin, J., and Hutchings, B. (1999). A reconfigurable arithmetic array for multimedia applications. *International Symposium on Field Programmable Gate Arrays*, pages 135–143.

[MicroBlaze, 2010] MicroBlaze (2010). MicroBlaze Soft Processor Core. *Available at http://www.xilinx.com/tools/microblaze.htm.*

[Miyamoto and Ohmi, 2008] Miyamoto, N. and Ohmi, T. (2008). Delay Evaluation of 90nm CMOS Multi-Context FPGA with Shift-Register-type Temporal Communication Module for Large-Scale Circuit Emulation. *IEEE International Conference on Field-Programmable Technology (ICFPT)*, pages 365–368.

[Mrabet, 2009] Mrabet, H. (2009). Design and optimization of reconfigurable architectures: The FPGA Family. *PhD. Thesis : http://www-asim.lip6.fr/publications/.*

[NIOS, II] NIOS (II). NIOS II Processor. *available at http://www.altera.com/products/ip/processors/nios2/ni2-index.html.*

[Okamoto et al., 2004] Okamoto, T., Kimoto, T., and Maeda, N. (2004). Design Methodology and Tools for NEC Electronics Structured ASIC. *Proc. ISPD*, pages 90–96.

[OpenCores, 2010] OpenCores (2010). OpenCores Projects. *available at http://www.opencores.org.*

[Padalia et al., 2003] Padalia, K., Fung, R., Bourgeault, M., Egier, A., and Rose, J. (2003). Automatic transistor and physical design of FPGA tiles from an architectural specification. *FPGA*, pages 164–172.

[Peter Yiannacouras and Rose, 2007] Peter Yiannacouras, J. G. S. and Rose, J. (2007). Exploration and Customization of FPGA-Based Soft Processors. *IEEE Transactions on Computer Aided Design of Integrated Circuits and Systems*, 26(2):266–277.

[Phillips and Hauck, 2002] Phillips, S. and Hauck, S. (2002). Automatic layout of domain-specific reconfigurable subsystems for system-on-a-chip. *FPGA*, pages 165–173.

[Pistorius et al., 2007] Pistorius, J., Hutton, M., Schleicher, J., Iotov, M., Julias, E., and Tharmalignam, K. (2007). Equivalence Verification of FPGA and Structured ASIC Implementations. *FPL'07*, pages 423–428.

[S.Brown, 1994] S.Brown (1994). An Overview of Technology, Architecture and CAD Tools for Programmable Logic Devices. *Custom Integrated Circuits Conference*, pages 69–76.

[S.Dai and E.Bozorgzadeh, 2006] S.Dai and E.Bozorgzadeh (2006). CAD Tool for FPGAs with Embedded Hard Cores for Design Space Exploration of Future Architectures. *FCCM'06*, pages 329–330.

[Sherlekar, 2004] Sherlekar, D. (2004). Design considerations for Regular Fabrics. *Proc. ISPD*, pages 97–102.

[Sima et al., 2001] Sima, M., Cotofana, S., van Eijndhoven, J. T. J., Vassiliadis, S., and Vissers, K. (2001). An 8x8 IDCT Implementation on an FPGA-Augmented TriMedia. *Proceedings of the 9th Annual IEEE Symposium on Field-Programmable Custom Computing Machines*, pages 160–169.

[SKILL, 2010] SKILL (2010). SKILL Programming Language. *http://www.cadence.com.*

[S.Kirkpatrick et al., 1983] S.Kirkpatrick, C.D.Gelatt, and M.P.Vecchi (1983). Optimization by Simulated Annealing. *Science 220*, pages 671–680.

[STMicroelectronics, 2010] STMicroelectronics (2010). *http://www.st.com.*

[Stratix, IV] Stratix (IV). Stratix IV FPGAs, Device Handbook. *Available at http://www.altera.com/products/devices/stratix-fpgas/stratix-iv/literature/stiv-literature.jsp.*

[S.Yang, 1991] S.Yang (1991). Logic Synthesis and optimization benchmarks, Version 3.0. *Microelectronics Center of North Carolina (MCNC), Raleigh.*

[Synopsys, 2010] Synopsys (2010). http://www.synopsys.com/home.aspx.

[Tabula, 2010] Tabula (2010). *http://www.tabula.com.*

[T.Cormen et al., 1990] T.Cormen, C.Leiserson, and R.Rivest (1990). Introduction to Algorithms. *MIT Press, Cambridge.*

[TIERLOGIC, 2010] TIERLOGIC (2010). *http://www.tierlogic.com.*

[Trimberger et al., 1997] Trimberger, S., Carberry, D., Johnson, A., and Wong, J. (1997). A Time-Multiplexed FPGA. pages 22–28.

[Underwood and Hemmert, 2004] Underwood, K. and Hemmert, K. (2004). Closing the gap: CPU and FPGA trends in sustainable floating-point BLAS performance. *Proc. FCCM*, pages 219–228.

[V.Betz and J.Rose, 1997] V.Betz and J.Rose (1997). VPR: A New Packing Placement and Routing Tool for FPGA Research. *International Workshop on FPGA*, pages 213–22.

[Verma and Akoglu, 2007] Verma, R. and Akoglu, A. (2007). A Coarse Grained Reconfigurable Architecture For Variable Block size Motion Estimation. *IEEE International Conference on Field-Programmable Technology (ICFPT)*, pages 81–88.

[Virtex, 5] Virtex (5). Virtex-5 Multi-Platform FPGA. *Available at http://www.xilinx.com/products/silicon_solutions/fpgas/virtex/virtex5/.*

[Wu and Tsai, 2004] Wu, K. and Tsai, Y. (2004). Structured ASIC, Evolution or Revolution. *Proc. ISPD*, pages 103–106.

[Xilinx, 2010] Xilinx (2010). Xilinx. *http://www.xilinx.com.*

[Ye and Rose, 2006] Ye, A. G. and Rose, J. (2006). Using Bus-Based Connections to Improve Field-Programmable Gate-Array Density for Implementing Datapath Circuits. *IEEE Transactions on Very Large Scale Integration (VLSI) Systems*, 14(5):462–473.

[Yu, 2007] Yu, C. (2007). A Tool for Exploring Hybrid FPGAs. *FPL*, pages 509–510.